AF847017

Atmosphere, Earth, Ocean & Space

Editor-in-Chief

Wing-Huen Ip, Institute of Astronomy, National Central University, Zhongli, Taoyuan, Taiwan

Series Editors

Masataka Ando, Center for Integrated Research and Education of Natural Hazards, Shizuoka University, Shizuoka, Japan

Chen-Tung Arthur Chen, Department of Oceanography, National Sun Yat-Sen University, Kaohsiung, Taiwan

Kaichang Di, Institute of Remote Sensing and Digital Earth Chinese Academy of Sciences, Beijing, China

Jianping Gan, Hong Kong University of Science and Technology, Hong Kong, China

Philip L.-F. Liu, Department of Civil and Environmental Engineering National University of Singapore, Singapore

Ching-Hua Lo, Department of Geosciences, National Taiwan University, Taipei, Taiwan

James A. Slavin, Department of Atmospheric, Oceanic and Space Sciences, University of Michigan, Ann Arbor, MI, USA

Keke Zhang, Space Science Institute, Macau University of Science and Technology, Macau, China

R. D. Deshpande, Geosciences Division, Physical Research Laboratory, Gujarat, India

A. J. Timothy Jull, Geosciences and Physics, University of Arizona AMS Laboratory, Tucson, AZ, USA

The series Atmosphere, Earth, Ocean & Space (AEONS) publishes state-of-art studies spanning all areas of Earth and Space Sciences. It aims to provide the academic communities with new theories, observations, analytical and experimental methods, and technical advances in related fields. The series includes monographs, edited volumes, lecture notes and professional books with high quality. The key topics in AEONS include but are not limited to: Aeronomy and ionospheric physics, Atmospheric sciences, Biogeosciences, Cryosphere sciences, Geochemistry, Geodesy, Geomagnetism, Environmental informatics, Hydrological sciences, Magnetospheric physics, Mineral physics, Natural hazards, Nonlinear geophysics, Ocean sciences, Seismology, Solar-terrestrial sciences, Tectonics and Volcanology.

More information about this series at http://www.springer.com/series/16015

Pao K. Wang

Motions of Ice Hydrometeors in the Atmosphere

Numerical Studies and Implications

Springer

Pao K. Wang
Research Center for Environmental Changes
Academia Sinica
Taipei, Taiwan

National Taiwan University
Taipei, Taiwan

University of Wisconsin-Madison
Madison, WI, USA

ISSN 2524-440X ISSN 2524-4418 (electronic)
Atmosphere, Earth, Ocean & Space
ISBN 978-981-33-4430-3 ISBN 978-981-33-4431-0 (eBook)
https://doi.org/10.1007/978-981-33-4431-0

© Springer Nature Singapore Pte Ltd. 2021
This work is subject to copyright. All rights are reserved by the Publisher, whether the whole or part of the material is concerned, specifically the rights of translation, reprinting, reuse of illustrations, recitation, broadcasting, reproduction on microfilms or in any other physical way, and transmission or information storage and retrieval, electronic adaptation, computer software, or by similar or dissimilar methodology now known or hereafter developed.
The use of general descriptive names, registered names, trademarks, service marks, etc. in this publication does not imply, even in the absence of a specific statement, that such names are exempt from the relevant protective laws and regulations and therefore free for general use.
The publisher, the authors and the editors are safe to assume that the advice and information in this book are believed to be true and accurate at the date of publication. Neither the publisher nor the authors or the editors give a warranty, expressed or implied, with respect to the material contained herein or for any errors or omissions that may have been made. The publisher remains neutral with regard to jurisdictional claims in published maps and institutional affiliations.

This Springer imprint is published by the registered company Springer Nature Singapore Pte Ltd.
The registered company address is: 152 Beach Road, #21-01/04 Gateway East, Singapore 189721, Singapore

Preface

This monograph is a summary of our recent research works in the study of motions (mainly the fall attitudes) of ice hydrometeors. "Ice hydrometeors" is a term used by atmospheric scientists to describe ice particles formed in atmospheric clouds such as cloud ice crystals, snow crystals, snowflakes, graupel, and hailstones. These are relatively large particles except the cloud ice crystals, and hence their hydrodynamic behaviors tend to be more complicated than small particles such as cloud droplets and small ice crystals. This imposes a serious problem for theoretical treatment on their motions, and the traditional analytical methods can do very little in solving the related problems. Numerical methods are obviously needed. But the computing technology (both hardware and software) up to the 1990s was just not advanced enough to afford easy solutions for most researchers.

But things started to change in the late 1990s and early 2000s as the advances in computing technology made the study of motions of complex-shaped bodies possible and affordable. In my own group, the first success was achieved only in 2012 when we obtained a set of good solutions for the three-dimensional flow fields around falling conical graupel using the CFD package Fluent, and the fields look reasonable when compared with some experimental data [5, 7]. But at that stage, the fall attitude of the graupel is fixed so that it is not really falling freely. The next feat was achieved in 2014–2015 when we utilized the 6-degrees of freedom and dynamic mesh features of Fluent and successfully obtained the flow fields around freely falling hexagonal ice plates. The plates are allowed to respond to the flow field so they can change positions as well as their fall attitude. They can rotate, zigzag, spiral and swing as the circumstantial requires. This was published in Cheng et al. [1], and we believe that is the first successful simulation of the completely free fall motion of ice crystals. It is fair to say that this opened a new era for the research of ice particle hydrodynamics in cloud physics. We demonstrated that this can be done—the technology is here! And all one needs in the future is to secure adequate computing resource and apply similar techniques to perform this type of studies. Soon after, we worked out the problem for freely falling hexagonal ice columns [4] and freely falling conical graupel [2, 3]. Both simulated columns and conical graupel perform tumbling motion at appropriate Reynolds number range in addition to other complex fall attitudes. More recently, we worked out the cases for several planar ice crystals [6]. We are now working on the snow aggregate problem.

Most of the materials presented here are taken from our published journal papers with most figures redrawn to suit the monograph better. In this monograph, we describe not only the hydrodynamic behaviors but also the ventilation effect of ice hydrometeors due to their motions. The ventilation effect is important in the diffusion growth and dissipation (evaporation) of hydrometeors.

I would like to take this opportunity to thank my former students and colleagues. I have been very fortunate to have this group of talented and dedicated researchers working with me. Without them, these works are impossible. Colleagues involved in this research are: Mr. Alexander Kubicek, Dr. Kai-Yuan Cheng, Prof. Tempei Hashino, Prof. Chih-Che Chueh, Mr. Joseph Nettesheim, and Mr. Jobst Muesser. During my tenure as Professor of Atmospheric and Oceanic Sciences in the University of Wisconsin-Madison, my research in cloud physics was mainly supported by US National Science Foundation (NSF). During my terms as the Director of Research Center for Environmental Changes (RCEC), I also received generous support by Academia Sinica (AS), Taiwan, particularly from Professor Chi-Huey Wong, former President of AS, who recruited me and provided strong support to RCEC, and Prof. Yuan-Tse Lee, former AS President previous to Wong and 1986 Nobel Laureate in Chemistry, who inspired me on the importance of global sustainability. In addition, I would like to thank the generous support of my research by US National Science Foundation grants ATM-0729898, AGS-1219586, and AGS-1633921; Taiwan Academia Sinica grant AS-TP-107-M10; and Ministry of Sciences and Technology grant MOST 109-2111-M-001-001. The book project was initiated with the encouragement of Academician Wing-Huen Ip.

Taipei, Taiwan
September 2020

Pao K. Wang

References

1. Cheng KY, Wang PK, Hashino T (2015) A numerical study on the attitudes and aerodynamics of freely falling hexagonal ice plates. J Atmos Sci 72:3685–3698
2. Chueh CC, Wang PK, Hashino T (2017) A preliminary numerical study on the time-varying fall attitudes and aerodynamics of freely falling conical graupel particles. Atmos Res 183:58–72
3. Chueh C-C, Wang PK, Hashino T (2018) Numerical study of motion of falling conical graupel. Atmos Res 199:82–92
4. Hashino T, Cheng K-Y, Chueh C-C, Wang PK (2016) Numerical study of motion and stability of falling columnar crystals. J Atmos Sci 73:1923–1942
5. Kubicek A, Wang PK (2012) A numerical study of the flow fields around a typical conical graupel falling at various inclination angles. Atmos Res 118:15–26
6. Nettesheim J, Wang PK (2018) A numerical study on the aerodynamics of freely falling planar ice crystals. J Atmos Sci 75:2849–2865
7. Wang PK, Kubicek A (2013) Flow fields of graupel falling in air. Atmos Res 124:158–169

Contents

Chapter 1
Clouds and Precipitation Particles

1.1 Clouds in the Atmosphere

Clouds are a common sight in the sky that is familiar to most people on earth. During a bright sunny day, we enjoy the sight of white puffy cumulus clouds floating, seemingly effortlessly, in the blue sky (Fig. 1.1).

But science teaches us that the "effortless floating" is actually a balance between the air updraft and the gravity—the gravitational force pulls the cloud drops—the constituent particles of small cumulus clouds—downward while the updraft force pushes the drops upward. When the two forces are about equal, the cloud, which is just an ensemble of these drops, appears floating in the sky to human eyes. This is to say that, in reality, the drops are falling relative to the air.

The same happens to other types of clouds. Take, for example, the hair-like cirrus clouds (Fig. 1.2).

Cirrus clouds usually occur at a very high altitude (higher than 6000 m) where the temperature is very much below the freezing point of water. Hence they are usually thought to consist nearly all ice, and for thin cirrus, the ice particles are largely simple ice crystals instead of snow aggregates (more about the distinction of them later). Cirrus clouds are often (though not always) seen moving rapidly with upper level winds which indicates that the ice crystals in such clouds not only fall relative to the updraft (as they must do) but also move relative to air horizontally as ice crystals are massive and hence must move in a different manner than air molecules.

Unlike small cloud droplets that are very nearly perfect spheres, ice crystals have multitude of complex shapes [5, 12]. Even the simplest of them, hexagonal columns and plates, are highly nonspherical and hence their orientation will have strong impact on their many physical properties.

For example, Fig. 1.3 shows an optical phenomenon called halo. This occurs when a type of cirrus clouds, called cirrostratus, covers the part of sky where the sun is located. A bright circle with color (red inside and blue outside) is seen around the sun. We now know that the halo is due to the refraction of sunlight by ice crystals in cirrostratus. Depending on the kind of ice crystals present in the cloud and the

© Springer Nature Singapore Pte Ltd. 2021

P. K. Wang, *Motions of Ice Hydrometeors in the Atmosphere*, Atmosphere, Earth, Ocean & Space, https://doi.org/10.1007/978-981-33-4431-0_1

Fig. 1.1 Fair weather cumulus clouds (Taipei, Taiwan)

Fig. 1.2 Cirrus clouds (Madison, Wisconsin, USA)

Fig. 1.3 Sun halo over cirrostratus cloud (Madison, Wisconsin, USA)

path of the sun ray, the angle subtended by the halo circle can be 22° or 46°. But the calculation of the halo angles assumes that the ice crystals (assumed to be either hexagonal plates or columns) are oriented with largest dimension horizontal. To assume such an orientation during the fall, the ice crystals have to reach a certain size. Otherwise the orientation can be random and we won't be able to see the halo. This indicates that the optical properties of the cloud can depend on the motion of the ice crystals.

When the crystals reach an even larger size, their fall attitude may become oscillatory. When this happens the ice crystal orientation will change which, in turn, will change the ray path of the sunlight shinning on them. This oscillation can be observed as we sometimes see another optical phenomenon associated with cirrostratus called circumzenithal arc (Fig. 1.4). Theoretical calculations by McDowell [9] verified that this arc can be seen most frequently when the ice plates oscillate with a tipping angle of 1°. Very small angle but large impact on the optics! This underlines the importance of particle motions.

Ice crystals in cirrus-type clouds are usually small, often a few tens to ~100 μm. There are clouds that contain much larger ice particles such as snowflakes, graupel and hailstones. The type of clouds that contain all these kinds of particles is the cumulonimbus (usually abbreviated as Cb by meteorologists), also known as thundercloud, and Fig. 1.5 shows an example of such cloud type. Strong winds and heavy rain, even large hailstones, often accompany the presence of such clouds.

Fig. 1.4 Circumzenithal arc (Madison, Wisconsin, USA)

There are many other types of clouds and they are all ensembles of cloud and precipitation particles. These particles are all suspended in a viscous medium—air, and we call such a system **colloid**. There are sometimes misunderstandings about the nature of such a system, and the most common one is probably the relation between particles in a colloid. Particles in a colloid certainly can interact with each other, for example, by collision or by competing for water vapor for growth. But such interactions usually occur at very short range, i.e., at a distance much smaller than the dimension of the particles. Otherwise, the particles behave as completely independent from each other. This is quite different from a system of tightly bounded particles such as an atom or a molecule which is bounded by forces much stronger than ordinary forces in the natural environment. Take the simplest of atoms, the hydrogen atom, as an example. It consists of a proton (with positive electric charge) and an electron (with negative electric charge), and the two are bounded together by a long range coulomb force, and it takes a lot of energy, called the **ionization energy** (2.18×10^{-18} J or 13.6 eV), to break apart this bound. Only the so-called ionizing radiation, such as X-rays or gamma rays, which are rare near the earth surface, can cause the ionization of a hydrogen atom. In contrast, the particles in a cloud are hardly bounded with each other, each moves nearly independently. When a hydrogen atom moves, the proton and electron in it move together as a unit because the two is strongly bounded. When a cloud moves, the particles in it move independently. The cloud appears to move as an entity but it is really due to the motion of the air, i.e.,

Fig. 1.5 Cumulonimbus (thundercloud) (Prague, Czech Republic)

winds or waves, that carries these particles together forward. This means that the particles mainly respond to the atmospheric environmental conditions rather than to the condition of their fellow particles.

Thus, aside from the occasion of collision between particles, we can safely assume that the particles move individually without worrying about the motions of other particles most of the time.

1.2 Impact of Particle Motions on the Physics of Clouds

This monograph is about the study of the motions of ice particles in clouds. Before we delve into the subject of particle motions, let's briefly examine why it is important to study particle motions in clouds.

First of all, we have said that cloud drops float because their weights are supported by the updraft of air. If the drops become larger such that the updraft can no longer support their weight, these drops will fall and when they fall to or near the ground, they become raindrops. The same applies to other types of cloud particles, that is, when cloud particles become too large to be supported by updraft, they become **precipitation** particles. This implies that the fall speed of these particles (collectively called **hydrometeors**), is of importance in deciding when a cloud precipitates. Once most of the cloud moisture precipitates, the cloud dissipates to either disappear completely or become a small remnant.

But how do cloud particles become larger? They have to grow by some physical processes. Hydrometeors grow by two main modes: diffusion growth and collision growth. Diffusion grow is the mode via which water vapor is transported toward the particle surface such that the particle becomes larger. The opposite of this—water vapor is transported away from the particle surface—is called evaporation (sublimation in the case of ice turning into vapor). Collision growth is the mode via which particle become larger by colliding and coalescing with other particles, and the opposite of this process is the breaking of the particle which makes the particle smaller. Both diffusion and collision growth are influenced by the motion of the hydrometeor.

Some growth processes also involve phase change of water. Diffusion growth and evaporation both are phase change and hence involve the release and consumption of latent heats that will influence the local thermodynamics of the cloud region where these hydrometeors are located. Collision and subsequent freezing of supercooled water drops with an ice particle, a process called riming which is the main growth mode of graupel and hail, also involves phase change and hence the release of latent heat. Since these growth processes are influenced by the particle motion, it follows that local cloud thermodynamics must also be influenced by particle motions as well.

1.3 Cloud and Precipitation Particles

We have mentioned the cloud and precipitation particles brief before. We will now give a more precise description about them in the following. The names used in the scientific community evolve from time to time, so readers should not be surprised if some of terms appear to be different from common usage or even terms from previous scientific discussions.

Only brief summary of the main characteristics of these particles will be discussed here. Detailed properties of them will be discussed in later chapters when their hydrodynamics are examined.

1.3.1 Cloud Particles

Cloud drops—these are small water droplets from a few to a hundred μm. The typical radius of cloud drops is 10 μm. At such small size range, the drops are nearly perfect spheres. Since they are small, their fall velocities are also small. Cloud particles can fall at any velocity relative to air from 0 to the "terminal velocity" u_∞. A particle reaches the terminal velocity when the gravitational force is balanced by the drag force and this will be discussed in Sect. 3.4. Typical cloud droplets of a few tens of μm radius fall at terminal velocities of a few cm/s in regular cloud condition (see [13]).

Cloud ice—also called "ice crystals" or "pristine ice". These are supposedly small, single ice crystals. The typical sizes range from a few tens to a few hundred μm. There are some observations show that some cloud ice are not of complete shape but are fragments, although there is a possibility that fragmentation is caused by the sampling process. Ice crystals are rather fragile and may collide with the sampler inlet and break.

Ice crystals can have many different "habits" (the shape of the crystal form), e.g., hexagonal columns, hexagonal plates, bullet rosettes, dendrites, broad branches, etc. However, most of the ice crystal samples collected in situ in clouds are of simpler shapes and rarely of more elaborated shapes.

There are two major axes of the crystallographic structure of ice, namely, *c*- and *a*-axis, as shown in Fig. 1.6.

As shown in this figure, the *c*-axis is the one perpendicular to the basal plane (the shaded hexagon) while the *a*-axis is the one parallels to the basal plane. If an ice crystal is longer in the *c*-axis direction, it becomes a columnar crystal (the ice crystal on the left). On the other hand, if an ice crystal has larger dimension along the *a*-axis, it becomes a planar ice crystal (the crystal on the right). This is the case of single crystals. There are polycrystalline ice crystals whose axes are much more complicated and more details can be found in Hobbs [3] and Pruppacher and Klett [10].

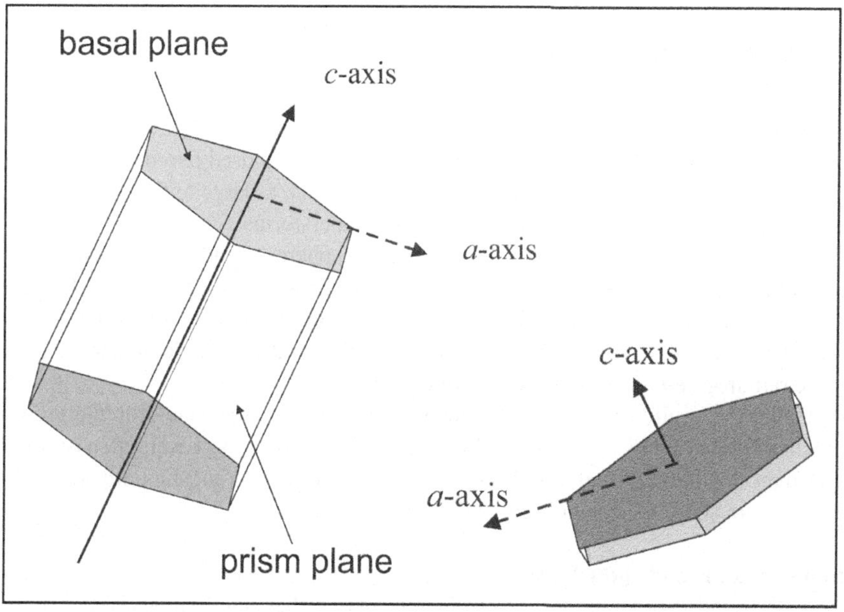

Fig. 1.6 The *c*- and *a*-axis of ice crystals

Not all ice crystals can be classified so neatly as columnar type or planar type but there are many other complicated shapes that require special terminology to deal with. The descriptions of the shapes are important as each type exists in certain favorable environment and hence the detection of them may indicate the environmental conditions. This may be useful for remote sensing of the atmosphere. In addition, the shape of an ice crystal impacts the optical properties of the crystals which may, in turn, impact the climate because of the influence on solar and terrestrial radiation by ice clouds (e.g., [6]).

Bailey and Hallett [1] summarized the favorable environmental conditions for certain cloud ice crystal types. The two most important factors determining the ice crystal habits are the temperature and saturation condition. Readers are referred to that paper or see Wang [12, Chap. 9, p. 249] for details.

Rimed ice crystals—some ice crystals have "rime" on them. The rimes are due to the spontaneous freezing of supercooled cloud drops when they collide with an ice particle. The name "supercooled" means that these drops have surface temperature below 0 °C and yet remain as liquid. Supercooled drops and ice particles cannot be in stable equilibrium due to their difference in saturation vapor pressure at the same temperature (see [10] or [12] for an explanation).

Rimed ice crystals may be regarded as an intermediate particle type between the cloud particles and precipitation particles. Lightly rimed crystals may remain largely in clouds whereas heavily rime crystals can be massive enough to fall and become "precipitation" particles.

1.3.2 Precipitation Particles

Rain drops—these are large water drops that updraft can no longer support and fall from clouds. Rain drops are usually larger than a few hundred μm in diameter and the typical diameters is 1 mm (i.e., 1000 μm). They can as large as 5–7 mm in many heavy rain events especially in warm season thunderstorms.

Unlike cloud drops which are close to perfect spheres, raindrops are distorted to nonspherical shapes and the degree of distortion depends on drop size, fall speed and other environmental conditions. Figure 1.7 shows two large rain drops (largest dimension 6.4 and 8.4 mm) photographed by high speed camera demonstrating that large raindrops are far from spherical during their fall.

Under the equilibrium, non-oscillatory condition, the larger the drop the greater the distortion [12, Fig. 2.13]. Natural falling large raindrops, however, often oscillate and may not fall vertically, and their shapes often deviate substantially from the equilibrium shape as shown in Fig. 1.7.

Snow crystals and snowflakes

The precipitating particles during a snowfall can be either single snow crystals (often mm-sized or larger) or aggregate of snow crystals, the latter are commonly known as snowflakes. They are also called snow aggregates. They form by the collision and

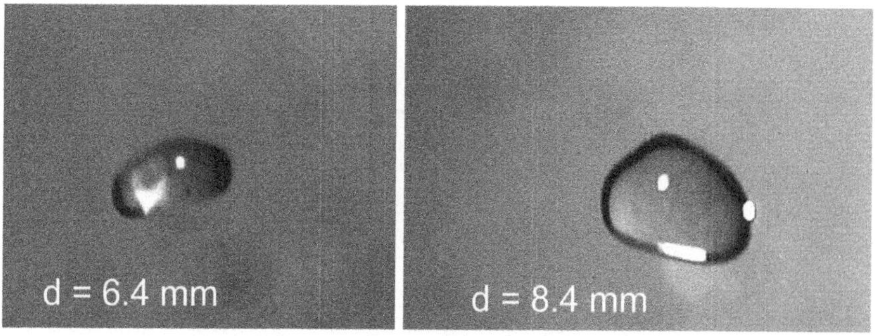

Fig. 1.7 Two examples of falling raindrops taken by a high-speed camera. Not all drops are falling at their equilibrium shapes

coalescence of ice crystals and the number of individual crystals can range from a few to a few dozens. Figures 1.8 and 1.9 show two examples of snow crystals and snowflakes deposited on a pad on the ground and the window of a commercial jet parking in an airport. The snow crystals in the first example are relatively large

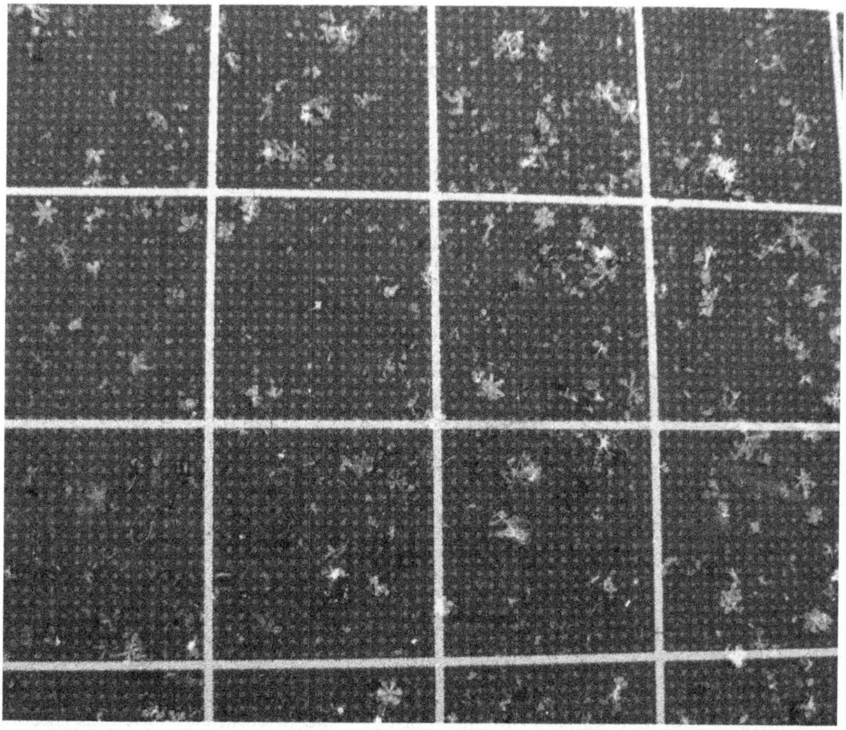

Fig. 1.8 An example of snow crystals falling on a pad. The white grid of the pad is 1″ × 1″ (Madison, Wisconsin, USA)

Fig. 1.9 A sample of snow crystals falling on the window of a commercial jet (Madison, Wisconsin, USA)

and that in the second are relatively small. Currently, little is known exactly how ice crystals stick together to form snowflakes, much less about their collision and coalescence efficiency (and hence the growth rates). One motivation of the research whose results are reported in this monograph is to understand the fall behavior of ice particles, especially the ice and snow crystals and some snowflakes, such that our knowledge on the collision behavior between ice particles can be advanced.

Another example of snowflakes is shown in Fig. 1.10. These are relatively large flakes, each individual crystal can be as large as 6–7 mm and the whole flake can be several centimeters in dimension consisting of several to a dozen individuals. Such large flakes are quite common during a blizzard and their wiggling zigzag motions are the stereotypical image of winter or Christmas in northern countries.

Sleet

In the United States, the term sleet refers to ice pellets which are solid ice particles composed of frozen or mostly frozen raindrops or refrozen partially melted snowflakes according NOAA glossary (https://w1.weather.gov/glossary/index.php?word=sleet). These pellets of ice usually bounce after hitting the ground or other hard surfaces and also make a fine, high pitch sound in contrast to the near silence of falling snowflakes or the splat of raindrops. Heavy sleet that accumulates to more than 0.5″ is a relatively rare event.

Fig. 1.10 Snowflakes falling on the hood of a car. Large flakes can consist of a few dozens of individual dendrites (Madison, Wisconsin, USA)

Graupel

When riming of ice crystals goes on for a considerable amount of time, the rime eventually covers the entire crystal which becomes unrecognizable at this stage. The particle now becomes a graupel, sometimes called a soft hail. They are typically conical in shape, more or less similar to that of a wax apple, and some examples are shown in Fig. 1.10. They can be of other shapes as well [4].

Graupel are the predecessor of hailstones. In cloud physics, graupel is defines as such rimed particles with diameter less than 5 mm. Further riming of graupel will eventually lead to the formation of hailstones.

Takahashi [11] proposed that the collision between graupel and ice crystals in cloud is the most efficient charging mechanism in thunderclouds, and many recently studies (e.g., [7, 8]) have confirmed that that lightning in clouds often collocated with the layer laden with large concentration of graupel, making Takahashi's theory highly likely to be true.

Hailstones

As already explained above, hailstones are the products of riming also. When a graupel continues to grow by riming to a size greater than 5 mm, it becomes a hailstone. Hailstones can be very large and can be golf ball or even baseball size in intense thunderstorms. Figures 1.12 shows a sample of the hailstones fallen in the backyard of the author's house in Madison, Wisconsin, USA, on April 13, 2006. These are 3–4 cm diameter stones, i.e., about that of a golf ball.

The largest hailstone on record is about 20 cm (8 in.) in diameter and weights nearly 0.88 kg [12].

Hailstones are usually spheroidal in shape but with different texture on the surface. Some are relatively smooth and some have coarse rimes on them. Some hailstones grow large lobes.

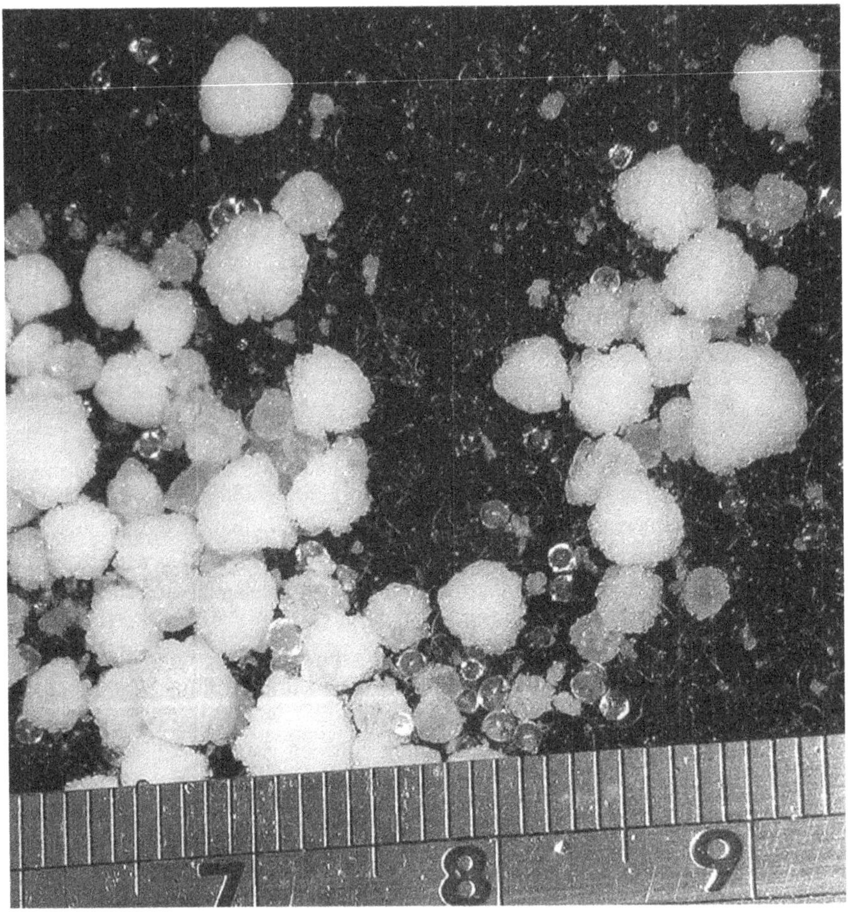

Fig. 1.11 A sample of conical graupel (Madison, Wisconsin, USA)

Fig. 1.12 A sample of hailstones (Madison, Wisconsin, USA)

Because of their size, large hailstones can cause severe damages to properties and crops and injuries and death to lives. Every year, the loss of agricultural yields reaches billions of US dollars per year due to falling hail in severe storms. Evaporation of hailstones falling through subsaturated air beneath the cloud causes rapid cooling of the ambient air which can sink rapidly since colder air is also heavier at the same pressure. This rapidly sinking air can cause the so-called "downburst" phenomenon in air near the surface. The downburst phenomenon is believed to be the cause of some air crashes as an aircraft flying into such a strongly sinking air would have difficulty pulling the nose upward necessary for either landing or taking off [2].

References

1. Bailey MP, Hallett J (2009) A comprehensive habit diagram for atmospheric ice crystals: confirmation from the laboratory, AIRS II, and other field studies. J Atmos Sci 66:2888–2899
2. Fujita TT (1978) Manual of downburst identification for Project Nimrod. SMRP research paper. Satellite and mesometeorology research project. Dept. of the Geophysical Sciences, University of Chicago, Chicago, 104 pp
3. Hobbs PV (1974) Ice physics. Oxford University Press, 837 pp
4. Knight CA, Ashworth T, Knight NC (1978) Cylindrical ice accretions as simulations of hail growth. Part II: the structure of fresh and annealed accretions. J Atmos Sci 35:1997–2009
5. Libbrecht K (2007) The snowflake. Voyageur Press, 156 pp
6. Liou KN (2002) An introduction to atmospheric radiation, 2nd edn. Academic Press, 583 pp
7. MacGorman DR, Straka JM, Ziegler CL (2001) A lightning parameterization for numerical cloud models. J Appl Meteor 40:459–478

8. Mansell ER, MacGorman DR, Ziegler CL, Straka JM (2005) Charge structure and lightning sensitivity in a simulated multicell thunderstorm. J Geophys Res 110:D12101. https://doi.org/10.1029/2004JD005287
9. McDowell RS (1979) Frequency analysis of the circumzenithal arc: evidence for the oscillation of ice-crystal plates in the upper atmosphere. J Opt Soc Am 69:1119–1122
10. Pruppacher HR, Klett JD (1997) Microphysics of clouds and precipitation, 2nd edn. D. Reidel, New York, p 954
11. Takahashi T (1978) Riming electrification as a charge generation mechanism in thunderstorms. J Atmos Sci 35:1536–1548
12. Wang PK (2013) Physics and dynamics of clouds and precipitation. Cambridge University Press, 467 pp
13. Wang PK, Pruppacher HR (1977) An experimental determination of the efficiency with which aerosol particles are collected by water drops in subsaturated air. J Atmos Sci 34:1664–1669

Chapter 2
Observational Studies of Ice Hydrometeors and Their Fall Behavior

In Chap. 1 we give a brief introduction on what ice hydrometeors are there in the atmosphere. In this chapter, we will discuss the scientific observations on the physical properties and hydrodynamic behavior of these ice hydrometeors and their hydrodynamic behaviors. Unsurprisingly, most observational studies were conducted at the ground level because of the ease to do so, but there were also aircraft and even satellite observations that examined the ice hydrometeors located high above in clouds. We make brief summaries of these observational studies below.

Ancient people did not have any chance to observe the fall behavior of cloud ice as there wasn't a way for human to ascent into clouds, but many certainly had had chances to see snowfall which often contains simple ice crystals such as hexagonal plates or dendrites and their fall behavior can be thought as the surrogate of those in clouds. In one of the well-known stories in Chinese literature, a famous Prime Minister Xie An (320–385 AD) of East Jin Dynasty was allegedly asking a group of nephews and nieces of his when snow began to fall: "How would you describe these fluttery snowflakes?" To this question, one of his nephews answered that "it's like salts spreading from the sky" whereas one of his nieces commented "it's rather like willow catkins blown up by winds". Xie An was reportedly happy with both descriptions. Indeed, both kinds of fall attitude can occur as high density ice particles (for example, solid columns and plates) do fall like salt grains whereas a dendritic aggregate fall like a willow catkin. In fact, a dendritic aggregate does have a shape similar to a catkin, and hence it's little surprise that they fall in similar attitudes!

Scientific systematic observations of the fall of ice particles began only in the twentieth century. Some of these studies were made by observing natural ice particles but there were also studies using artificial particles in laboratory. Natural ice particles are not always available for scientific study all the time. For ice and snow crystals, one has to be in a flurry or a blizzard, or in the best situation, in a mountain top laboratory immersed in ice clouds.

Japanese scientists appear to be one of the earliest group in making such observations. Fall velocity is one of the most important kinetic characteristics studied by

© Springer Nature Singapore Pte Ltd. 2021

P. K. Wang, *Motions of Ice Hydrometeors in the Atmosphere*, Atmosphere, Earth, Ocean & Space, https://doi.org/10.1007/978-981-33-4431-0_2

these investigations. Thus Nakaya and Terada [25] performed simultaneous observations of the mass, fall velocity and form of individual snow crystals. Later, Magono [21] measured the fall velocity of both single crystals and snow aggregates, and deduced that the fall velocity is approximately proportional to the square root of the maximum dimension of a snowflake. Magono [23] made further studies on the theoretical interpretation of the measured fall velocities of planar, dendritic, spatial and needle crystals, graupel, and snowflakes. Further studies on the fall velocity as well as fall attitude of snow were made by Litvinov [20], Bashkirova and Pershina [2], Magono and Nakamura [24], Fukuta [5], Brown [3], Jiusto and Bosworth [11], Heymsfield [6], and Jayaweera and Ryan [9].

Zikmunda and Vali [33] studied the fall patterns and fall velocities of rimed ice crystals of size between 0.02 and 0.5 cm using a stereoscopic camera system with stroboscopic illumination. They determined the relationships between terminal fall velocities and crystal dimensions for graupel, rimed columnar crystals, rimed plate crystals, rimed capped columns, rimed broken branches, and aggregates of rimed crystals.

A common opinion held by many scientists studying cloud and precipitation particles is that ice particles generally fall with maximum projected area perpendicular to the direction of fall. This may or may not be true. This opinion obviously stems from the observation that snowflakes usually fall with the flake oriented quasi-horizontally while falling quasi-vertically although they also tend to perform zigzag motion. That may be true for the planar ice crystals but there are many other type of crystals, e.g., columns, that can perform more complicated motions. Also, there are differences in fall pattern between rimed and unrimed crystals. And how about the fall pattern of conical graupel and hailstones that have lobes on their surface? These all need careful studies as they also impact the growth rates of these particles. Below we will just summarize a few more systematic studies.

One of the earliest systematic observation of the fall attitude of ice crystals, snowflakes and graupel was that of Magono [22]. A stroboscopic camera system was used to take photographs of the falling ice particles whose shape, fall speed and attitude were studied. The crystals were mostly dendritic or spatial dendritic crystals and very few needles. These dendritic crystals form snowflakes by collision. From the sounding of the day, Magono deduced that the crystals probably formed in the layer 2400–1800 m above ground while the snowflakes began to grow at a height ~1300 m.

Magono observed that the snowflakes have fall velocities somewhat larger than snow crystals, but for flakes larger than about 2 cm they seem to fall at almost the same velocity. Thus he theorized that the snowflakes are unlikely to grow by the collision between two flakes because they would keep the same vertical distance from each other and it would be difficult to collide. But since single snow crystals fall slower than flakes, the flake can grow by colliding with single crystals.

Graupel growth had been studied by Czermak [4], Barkow [1], Nakaya and Terada [25], and Sawada [30] who made either field observations or laboratory studies of the growth of graupel. Magono [22] also studied the graupel growth. Most of the graupel he studied are conical although there are some without an apex, presumably lost due to some (collision) processes. He observed that the graupel can fall in oblique orientation on its way down. He also speculated how the conical shape of graupel form.

Podzimek [27, 28] studied the fall of model snow crystals in water tank and determined the relationship between the drag coefficient and Reynolds number. The Reynolds number ranges in $10 < N_{\mathrm{Re}} < 19{,}000$. The upper limit is obviously much larger than natural snow crystals but may be relevant for hailstones. Based on this set of experiments, Podzimek [28] found that the drag coefficient C_D can be expressed as:

$$C_D = 16.5 N_{\mathrm{Re}}^{-0.468} \text{ (hexagonal plates with } N_{\mathrm{Re}} < 200) \tag{2.1}$$

$$C_D = 20.2 N_{\mathrm{Re}}^{-0.466} \text{ (stars with narrow or broad arms, plates with outgrows } N_{\mathrm{Re}} < 200) \tag{2.2}$$

Relation (2.1) was later verified by Jayaweera and Cottis [7].

Jayaweera and Mason [8] studied the fall behavior of cylinders and cones made of acrylic and metal materials of various size and shapes in viscous fluid tank for Reynolds number range from <0.01 to 1000, some of which can be used to represent ice particles from ice crystals up to large graupel and small hailstones in the atmosphere. Their finite cylinder cases have been discussed by Ji and Wang [10] and Wang and Ji [31] in some details. Ji and Wang's [10] numerical simulation nearly completely reproduced Jayaweera and Mason's [8] experimental results. Kubicek and Wang [14] and Wang and Kubicek [32] also simulated numerically the fall of conical bodies with fixed orientation and found that the results agree reasonably well with that of Jayaweera and Mason [8] despite the constraint of fixed orientation. In Chap. 4, we will discuss the detailed numerical simulation of the free fall of conical graupel in detail.

An extensive study on the fall attitude of conical particles was performed by List and Schemenauer [18] who performed laboratory observation on the free-fall behavior of man-made models of six planar snow crystals, two conical graupel and two conical small-hail particles in glycerin-water mixtures and salt solutions. These models were made of various materials. A set of four conical models, with spherical segments for bases, two representing conical graupel and two representing small hail particles were machined from perspex and polished. Three of the cones had apex angles of 90° and the fourth 70. The non-translatory motions of the conical models can be split up into two types of roughly periodic oscillations. The first is a sideways oscillation of the center of mass with a frequency ~0.2 Hz and amplitudes up to 1–2 particle diameters, depending on the model and Reynolds number. The second type is an angular oscillation or swinging of the axis of rotational symmetry with a frequency from 0.1 to 1 Hz.

List and Schemenauer's [18] results for conical particles are summarized in Fig. 2.1. The results were arbitrarily divided into five ranges of oscillations of the axis from stable to tumbling, with the boundaries between ranges accurate to about $\pm 10\%$

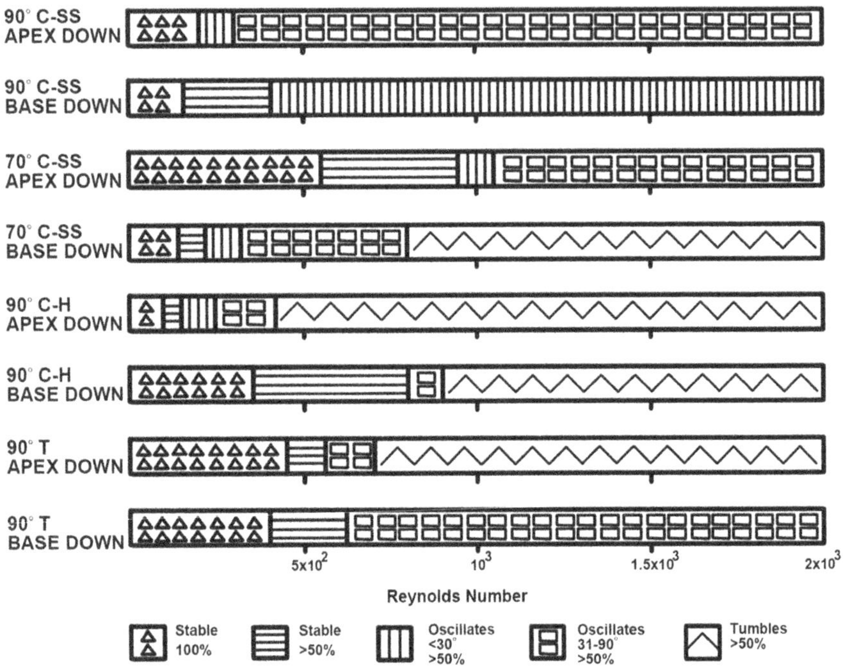

Fig. 2.1 Degree of oscillation of conical models as a function of Reynolds number: C-SS cone-spherical sector, C-H cone-hemisphere, T teardrop (after [18])

of the Reynolds number. From five to ten tests were made at a given Reynolds number, the type of hatching indicating the most frequent degree of oscillation observed. They found that the 90° teardrop shaped model is falling stably in a sideways position, independent of the position of release. Nevertheless, tumbling is reported at higher Reynolds numbers when released in the apex down position, because the teardrop model carries out one or more complete rotations to reach that stable sideways position. This peculiar alignment of the particle axis could explain why an ellipsoidal hailstone often contains a conical embryo particle with its main axis oriented at a right angle to the hailstone's stable fall direction, a direction parallel to its smallest axis. The 90° cone-spherical sector is more stable falling in the base down orientation, while the 70° cone-spherical sector is more stable falling in the apex down orientation. The latter represents the most stable combination of model and position for $Re < 1000$. The former combination, however, never shows oscillations >30° with an occurrence of more than 50%, even at $Re > 1000$. The 90° cone-hemisphere is more stable falling in the base down orientation. A discussion of the origin of the oscillations cannot be carried out here, as these secondary motions will be connected with the still unknown shedding of vortices in the wakes of the models or special boundary effects. In addition, it has to be pointed out that the stability situation for atmospheric particles may also be affected by a nonhomogeneous density distribution within single hydrometeors.

Zikmunda and Vali [33] mentioned previously studied the fall patterns of various ice particles of size 0.2–5 mm and provides rather detailed descriptions. Their findings are summarized below:

1. Planar crystals

   - The rimed plates were found to fall with the crystal planes oriented horizontally. No oscillations or side motions were observed for these crystals in agreement with laboratory results by Podzimek [27–29] for the range of Reynolds number 50–90. Thus they deduced that the accretion of supercooled droplets by the crystal does not change its fall attitude.
   - They also observed a few planar branch type crystals which are often rimed asymmetrically and hence the motions are more complicated. Spiraling and tumbling are possible.

2. Columnar ice crystals

   - All those crystals having axis ratios <3 fell steadily and with their major axes oriented horizontally, while only one crystal with an axis ratio >3 fell in a stable position and without rotation.
   - Most crystals exhibited several different types of rotatory motions: rotations occurred in both the vertical and horizontal planes about the minor axes of the crystal, and in at least one case a rotation about the major axis was also detected.
   - Almost all of the crystals exhibited slight drifts from the vertical during fall. The duration of observation for an individual crystal was small compared to the periods of the rotatory motions so that the complete cycles of the various motions can only be inferred from a grouping of all the observations.
   - The majority of the crystals deviate only slightly from the horizontal orientation but a few crystals exhibit very large deviations. Two out of the 23 crystals observed in detail were found to fall with their major axes close to the vertical.
   - The average angular velocity of rotation in the vertical plane was 14 rad s^{-1}, the highest values measured being ~45 rad s^{-1}

3. Capped columns

   - The rimed capped columns were found to fall steadily with their major axes (c axis of column) oriented either vertically or horizontally.
   - The orientation of the major axis appears to depend on the ratio of the plate diameter D to the column length L. For ratios of D/L larger than unity, the crystals fell like plates, i.e., with the column axis vertical in agreement with laboratory studies by Podzimek [29].
   - Riming was apparent on the outer surfaces of both plates but heavier on the bottom plate.
   - Accretion of supercooled droplets in the wake seems to have been very significant.
   - No riming occurred on the prism faces of these crystals thus tumbling was absent.

- Crystals with D/L < 1 fell with their major axes oriented horizontally but the presence of rimed droplets on the plate ends as well as on the prism faces can be taken as evidence for changes of orientation during fall.

4. Graupel

- Lump graupel falls mostly in a steady manner whereas conical graupel most frequently exhibits oscillations of the major axis about the vertical.
- The mean position of the apex of conical graupel is upward consistent with the opinion of List [16] but is in contradiction to Magono [22] who held that oblique position of the major axis was the stable fall attitude.

Kajikawa [12] measured the fall velocity of individual unrimed snow crystals. His results show that the fall velocity of ice crystals varies with crystal habit and apparently the higher density ones (such as thick plates and regular plates) fall much faster than lower density ones (such as dendrites, stellar crystals) for a given size which conforms with our common sense but in a quantitative way. Such quantitative observations are in fact quite important in cloud modeling if the cloud being modeled has ice crystals in it.

Pflaum et al. [26] performed a vertical wind tunnel study on the hydrodynamic behavior of growing, freely falling graupel. They grew the graupel from an initial frozen drop of 200–600 μm diameter suspended in the wind tunnel. Then small droplets produced from a steam machine were introduced into the wind tunnel which was chilled to below 0 °C temperature, thus the droplets in the steam were all supercooled. These droplets became rimes immediate upon collision with the frozen drop suspended in the tunnel beforehand so that the drop grew to become a graupel. The final size of the graupel so grown was 800 μm–1.2 mm in diameter. The supercooled droplets had diameter range 8–30 μm with a peak at 14 μm. They observed the following 5 basic fall modes:

(1) a rotational (spinning) motion around the ice particle's vertical axis;
(2) a helical translator motion;
(3) a 'bell-swing' or 'pendulum-swing' type of oscillatory motion;
(4) a precession of the axis of rotation;
(5) quiescent fall.

No tumbling motions, i.e. motions during which the ice particle would rotate by 360″ around the axis perpendicular to the fall direction, were observed. Note, however, that Pflaum et al. [26] only observed relatively small graupel. As we will see in our numerical studies of the graupel motion, tumbling will occur when graupel become larger.

Rimed ice particles which rotated about a vertical axis did so with 1–7 revolutions per second. The helical oscillation frequencies ranged from less than 1 Hz up to about 5 Hz. The radii of the helical paths ranged from less than one particle radius (tight helical motion) to greater than 10 particle radii. The frequency of the pendular oscillations ranged from 7 to 16 Hz. During such a pendular motion the angular excursion to either side of the axis of fall varied from 17 to 55″ and was not necessarily symmetric about the fall axis. If the axis of rotation precessed it did so with an angular excursion of 2–5″ about the vertical.

These fall modes appeared to be closely related to the instantaneous shape, surface roughness, and mass loading of the ice particle, all of which changed in time in a rather complicated manner.

Finally, the fall behavior of hailstones will be briefly summarized. Due to their large sizes (up to baseball and grapefruit size) and very high velocities (up to ~40 m s^{-1}), it is very difficult to perform laboratory study on their fall behavior because no vertical wind tunnel at present has vertical wind speed comparable to hailstones' fall speeds. It will also require a very long vertical distance for hailstones to reach terminal velocities and a very large facility would be needed if one is to conduct drop tower type experiments. Due to such difficulties, many hail fall attitude studies were made based on indirect inference of hail structure.

List [17] performed a study on the aerodynamics of hailstones based on ellipsoidal model hailstone experiments and found that they fall in the direction of the shortest axis and they fall stably—indicating that no tumbling in the fall. In addition, List et al. [19] found that falling spheroids can oscillate about a horizontal main axis and sometimes commence and carry out continuous rotations. The most general motion for a smooth spheroid turned out to be a gyration with the spin around the minor axis, whereby this minor axis follows the surfaces of a cone. Based on his observations, Lesins and List [15] estimated that the spin and nutation-precession frequencies of gyrating hailstones to be up to 50 Hz.

Knight and Knight [13], on the other hand, argued that moderate to large hailstones perform rapid symmetrical tumbling during their fall based on the study on the internal structure of the hailstones.

References

1. Barkow E (1908) Zur entstehung der graupeln. Meteor Z 25:456–458
2. Bashkirova TA, Pershina TA (1964) On the mass of snow crystals and their fall velocity. Tr GI Geofiz Observ Engl Transl 165:83–100
3. Brown SR (1970) Terminal velocities of ice crystals. M.S. thesis, Dept. of Atmospheric Science, Colorado State University, Fort Collins, 52 pp
4. Czermak P (1900) Zur Struktur und Form der Hagelkorner. Sitz Ber Wien II, a:185
5. Fukuta N (1969) Experimental studies on the growth of ice crystals. J Atmos Sci 26:522–531
6. Heymsfield A (1972) Ice crystal terminal velocities. J Atmos Sci 29:1348–1357
7. Jayaweera KOLF, Cottis RE (1969) Fall velocities of plate-like and columnar ice crystals. Q J R Meteor Soc 95:203–209
8. Jayaweera KOLF, Mason BJ (1965) The behavior of freely falling cylinders and cones in a viscous fluid. J Fluid Mech 22:709–720
9. Jayaweera KOLF, Ryan BF (1972) Terminal velocities of ice crystals. Q J R Meteor Soc 98:193–197
10. Ji W, Wang PK (1991) Numerical simulation of three-dimensional unsteady viscous flow past finite cylinders in an unbounded fluid at low intermediate Reynolds numbers. Theor Comput Fluid Dyn 3:43–59
11. Jiusto JE, Bosworth GE (1971) Fall velocity of snowflakes. J Appl Meteor 10:1352–1354

12. Kajikawa M (1972) Measurement of falling velocity of individual snow crystals. J Meteorol Soc Jpn Ser II 50:577–584
13. Knight CA, Knight NC (1970) The falling behavior of hailstones. J Atmos Sci 27:672–681
14. Kubicek A, Wang PK (2012) A numerical study of the flow fields around a typical conical graupel falling at various inclination angles. Atmos Res 118:15–26
15. Lesins GB, List R (1986) Sponginess and drop shedding of gyrating hailstones in a pressure-controlled wind tunnel. J Atmos Sci 43:2813–2825
16. List R (1958) Kennzeichen atmosphärischer Eispartikeln. I. Teil. Graupeln als Wachstumszentren von Hagelkörner. Z Angew Math Phys 9a:180–192
17. List R (1959) Der Hagelversuchskanal. Z Angew Math Phys 10:381–415
18. List R, Schemenauer RS (1971) Free-fall behavior of planar snow crystals, conical graupel and small hail. J Atmos Sci 28:110–115. https://doi.org/10.1175/1520-0469(1971)028%3c0110:FFBOPS%3e2.0.CO;2
19. List R, Rentsch UW, Byram AC, Lozowski EP (1973) On the aerodynamics of spheroidal hailstone models. J Atmos Sci 30:653–661
20. Litvinov IV (1956) Determination of falling velocity of snow particles. Izv Akad Nauk SSR Ser Geofiz Engl Transl 7:853–856
21. Magono C (1951) On the fall velocity of snowflakes. J Meteor 8:199–200
22. Magono C (1953) On the growth of snowflakes and graupel. Sci Rep Yokohama Natl Univ Sec I 2:18–40
23. Magono C (1954) On the fall velocity of solid precipitation elements. Sci Rep Yokohama Natl Univ Sec I 3:33–40
24. Magono C, Nakamura T (1965) Aerodynamic studies of falling snowflakes. J Meteor Soc Jpn 43:139–147
25. Nakaya U, Terada TJ (1935) Simultaneous observations of the mass, falling velocity and form of individual snow crystals. J Fac Sci Hokkaido Imp Univ Ser 2 1:191–200
26. Pflaum JC, Martin JJ, Pruppacher HR (1978) A wind tunnel investigation of the hydrodynamic behaviour of growing, freely falling graupel. Q J R Meteor Soc 104:179–187
27. Podzimek J (1965) Movement of ice particles in the atmosphere. In: Proceedings of the international conference on cloud physics, Tokyo and Sapporo, pp 224–230
28. Podzimek J (1968) Aerodynamic conditions of ice crystal aggregation. In: Proceedings of the international conference on cloud physics, Toronto. American Meteorological Society, pp 295–299
29. Podzimek J (1969) Contribution to the explanation of the motion of a falling column-like ice crystal. Stud Geophys Geod 13:199–206
30. Sawada T (1949) The aerodynamic investigation on the seed of Composita. Gakugei (J Hokkaido Gakugei Univ) 1(1):107–109
31. Wang PK, Ji W (1997) Numerical simulation of three-dimensional unsteady flow past ice crystals. J Atmos Sci 54:2261–2274
32. Wang PK, Kubicek A (2013) Flow fields of graupel falling in air. Atmos Res 124:158–169
33. Zikmunda J, Vali G (1972) Fall patterns and fall velocities of rimed ice crystals. J Atmos Sci 29:1334–1347

Chapter 3
Physics and Mathematics of the Hydrodynamics of Falling Ice Particles

3.1 Physical Configuration of the Problem

In this chapter, we will describe the theoretical methods in determining the free fall attitudes of ice hydrometeors of various shapes and the associated ventilation coefficient calculations. We will first describe the physical basis of the theoretical problems and then the mathematical techniques involved in attacking these problems.

For the fall of hydrometeors in clouds and the associated ventilation problem, the scale is small enough so that the compressibility of air plays no important role [11, 17]. Thus we only need to consider incompressible flow for the present study. For isothermal, incompressible flow of a Newtonian air fluid past an ice particle, the momentum and mass transport equations are:

$$\frac{\partial \vec{u}}{\partial t} + (\vec{u} \cdot \nabla)\vec{u} = -\frac{\nabla p}{\rho_a} + \nu \nabla^2 \vec{u} + \vec{g} \tag{3.1}$$

$$\nabla \cdot \vec{u} = 0 \tag{3.2}$$

where $\vec{u}$ stands for the air velocity vector, p the dynamic pressure, ρ_a the air density, ν the kinematic viscosity of air, and **g** the gravity. Equation (3.1) is the well-known time-dependent Navier–Stokes equation whereas Eq. (3.2) is derived from the continuity equation under the condition that ρ_a = constant for incompressibility assumption.

The general boundary conditions are

$$\vec{u} = 0 \text{ at ice particle surface} \tag{3.3}$$

$$\vec{u} = u_\infty \hat{e}_z \text{ sufficiently far away from the ice particle} \tag{3.4}$$

where u_∞ is the terminal fall velocity of the ice particle and $\hat{e}_z$ is an unit vector in the z-direction. The definition of the coordinate system used in this section is given

© Springer Nature Singapore Pte Ltd. 2021

P. K. Wang, *Motions of Ice Hydrometeors in the Atmosphere*, Atmosphere, Earth, Ocean & Space, https://doi.org/10.1007/978-981-33-4431-0_3

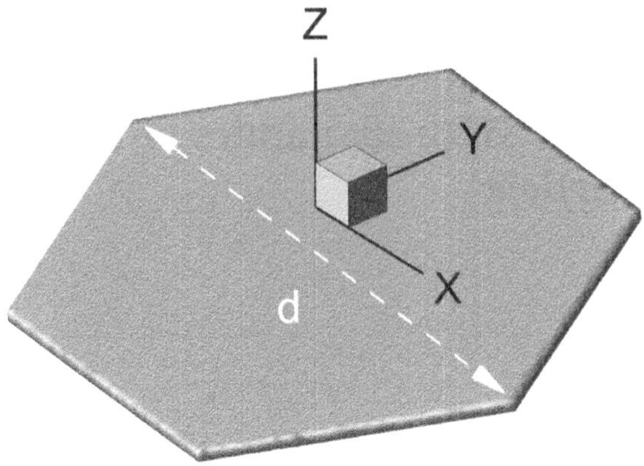

Fig. 3.1 The definition of the coordinate system using a hexagonal pate as an example

in Fig. 3.1. Equation (3.3), the 'inner boundary condition', is the nonslip condition indicating that the velocity on the ice surface is zero because it is a rigid solid whereas (3.4), the 'outer boundary condition', says that the air flow sufficiently far away is not disturbed by the presence of the ice particle as the mechanical influence of the particle motion should be of limited distance range only. In analytical methods, the 'sufficiently far away' criterion is usually replaced by the distance at infinity ($r \to \infty$) whereas in numerical methods one usually set the outer boundary at a distance many particle dimensions away for achieving the 'far way' effect.

Except some highly simplified cases, equation set (3.1)–(3.4) cannot be solved to obtain general analytical solutions and numerical methods are required to obtain solutions for most cases. Readers are referred to Pruppacher and Klett [11] or Wang [17] for a summary of the simple analytical solutions. The motions considered in this book are all complicated unsteady flows around freely falling ice hydrometeors and most (except spherical hailstones) are of complex shapes, and hence numerical methods are the only viable option for solution at present. In the case of spherical hailstones, the fall speeds are high and the wake flows are completely turbulent so that no analytical solutions have been found yet. Thus, the solutions mentioned in this monograph are all from numerical methods.

3.2 Numerical Methods of Solving Unsteady Incompressible Navier–Stokes Equation

Numerical solutions for flow past ice particles have been attempted by several investigators before and most of them are summarized in Pruppacher and Klett [11] and Wang [16]. Here it is sufficient to state that studies prior to 1989 were all of steady ($\partial \vec{u}/\partial t = 0$) axisymmetric flows. In addition, the shapes of ice particles were simplified to reduce the complexity. The simplifications of these previous numerical studies were made mainly because of the limitations of computing technology at the time in both hardware and software such that approximations were necessary to obtain any solution at all. For example, Pitter and Pruppacher [10] simulated the flow past hexagonal ice plates by assuming that hexagonal plate can be simplified as a thin oblate spheroid and the flow is steady so that the resulting flow field is steady state and symmetric with respect to the z-axis (though not symmetric in fore-aft direction as the standing eddies are simulated). Similarly, Schlamp et al. [12] simulated the flow past columnar ice crystals by assuming that ice columns can be approximated as infinitely long circular cylinders so that they possess symmetry along the center of the cylinder and the numerical problem is reduced to solving the 2-dimensional, time-independent Navier–Stokes equation pas a circle. While these studies contributed substantially to our initial understanding of the ice hydrometeor fluid dynamics, it is also obvious that such approximations have shortfalls and the results have serious limitations.

The first study to investigate unsteady flow past ice hydrometeors was due to Ji and Wang [6] where the shapes of hexagonal ice plates and broad branch ice crystals were treated as thin exact hexagonal plates and hexa-symmetric broad branch crystals, respectively. The flow was assumed to be unsteady so that the eddy shedding phenomenon could be simulated, and hence the exact time-dependent form of Eq. (3.1) was used. Similarly, Ji and Wang [7] studied the unsteady flow past an ice column by approximating hexagonal ice columns as finite-length circular cylinders. These two and a sequel by Wang and Ji [19] successfully simulated the 3-D (e.g., pyramidal and periodic vortex shedding) unsteady wake structure of falling ice plates, columns and broad-branch crystals and the eddy shedding phenomenon associated with such motions for the first time. The computed flow fields indeed match experimental measurements quite well in many aspects.

However, there is still a serious limitation in the studies mentioned in the paragraph above. In all three studies, the orientation and attitude of the ice particle is assumed to be fixed. This means that it is the ice crystal that determines how the air would move around it but the crystal would not respond to changes in the airflow. So, for example, an ice plate should feel the change in hydrodynamic drag when eddy shedding occurs and make adjustment of its orientation and posture (e.g., perform rotation, oscillation and tilting) in response to the change. The above assumption, however, prevents such adjustments to occur and the posture of the falling crystal remains unchanged. Also, the crystal fall velocity is also assumed to be constant, i.e., the flow field change would not impact the fall velocity of the ice crystal either.

This limitation may not be so serious when ice crystals are small as small crystals often perform quasi-steady fall attitude. For larger ice crystals, the assumption is becoming unrealistic, for example, it is well known that snowflakes during a blizzard perform complicated rotation, pendulum oscillation, zigzag fall path, lateral translation and tumbling. In addition, the fall velocity of the ice particle will change in response to the changes in the particle's fall attitude. None of these free fall behaviors can be simulated with the above assumption.

In order to remove this deficiency, the author and his research group members performed studies by allowing ice particles to perform totally free fall motion so that they can respond to changes in the flow field and make appropriate adjustments on their fall attitudes and velocities (and hence 3-D positions). We are fortunate to succeed in our attempts and the rest of this monograph is a summary of the results of these new studies.

3.2.1 *ANSYS Fluent*

As indicated above, we perform theoretical investigations on the fall behavior of freely falling ice particles by allowing the full interaction between the ice particle and the surrounding air such that the flow field of the air around the ice particle is disturbed by the particle motion and the change in the flow field feeds back to the particle to change the particle's velocity and posture. The particle-flow field interaction is very complicated and therefore numerical methods are necessary to solve such a problem. The numerical solver used for solving the Navier–Stokes equation in the works reported here is the commercial software ANSYS Fluent© versions from 13.1 to 16.2.0 (see https://www.ansys.com/products/fluids/ansys-fluent). The algorithm for solving the coupling of pressure and velocity fields is PISO (Pressure Implicit with Splitting of Operators, [5]), and the advection of momentum is solved with QUICK (quadratic upstream interpolation for convective kinetics, [2]). A second-order implicit scheme was chosen for transient formulation with the time step typically set to 1 or 2×10^{-5} s. The meshing method of Fluent is finite volume.

In this set of studies, we allow most ice particles to fall freely (except the initial study of conical graupel cases and large hailstones, the reason for the latter case will be stated later). This means that at each time step the particle not only can change their horizontal and vertical (x, y, z) positions but also their orientations (specified by the three Tait–Bryan angles-see discussions later in this chapter), hence there are totally six degrees of freedom in this kind of motions—3 degrees of freedom for the position (represented by the enter of mass of the particle) and another 3 degrees of freedom for the Tait–Bryan angles. This also implies that the numerical mesh used for the numerical calculation needs to be updated at each time step in order to achieve computational efficiency and accuracy. To achieve this purpose, we used the dynamic mesh with 6-degree of freedom (6DOF) capability of Fluent. One example of this mesh updating is shown in Fig. 3.2 (taken from [1]).

The dynamics-mesh option of Fluent, which smoothens and generates mesh based on mesh quality thresholds and lengths every few time steps. The movement of a falling ice particle is explicitly simulated based on the forces and torques acting on them. The computational domain was discretized mainly with tetrahedrons. Near the particle surface more (e.g., 10) layers of hexahedrons were implemented to better resolve the boundary layer (called inflation). We also use patch conforming grid generation and inflation in some cases. The domain normally consists of several hundred thousand to over a million cells, depending on the complexity of the case. The robustness of the results is confirmed with different numerical schemes and grid resolution (see [1] for a discussion). Therefore, we are confident that the motion and stability obtained in this study are robust results.

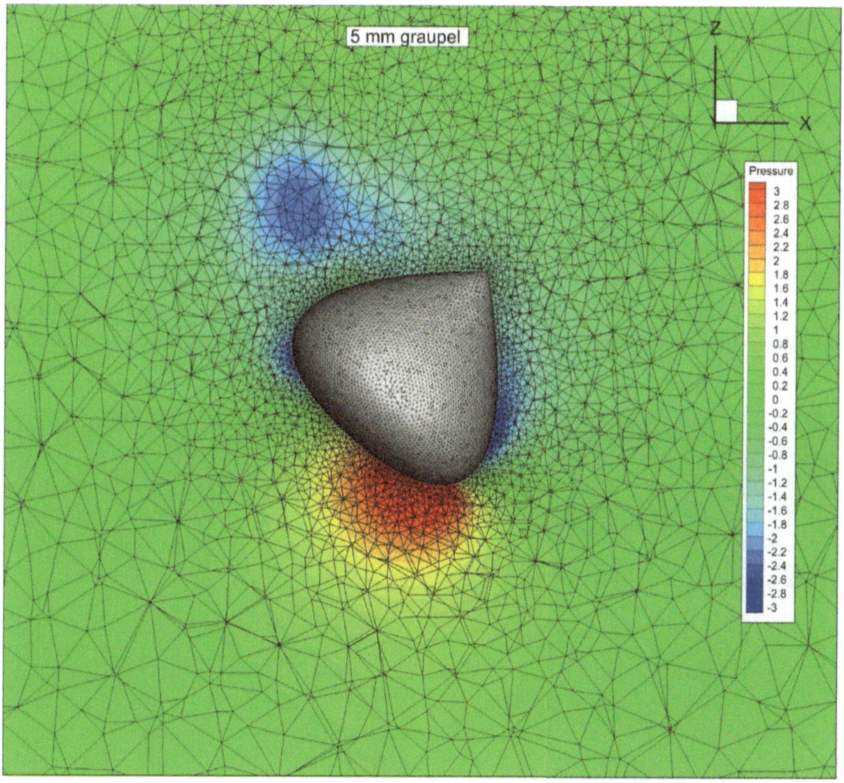

Fig. 3.2 The instantaneous numerical mesh for the case of a freely falling conical graupel of $d =$ 5 mm

For large spherical hailstones, we tested for a free fall case and found that its horizontal movement is very limited and much smaller than the vertical movement, i.e., it falls with little side motions. Thus we decided not to utilize the dynamic mesh capability for the hailstone cases.

3.2.2 Numerical Mesh Configuration

The design of numerical mesh including the specification of computational domains and the mesh techniques is an important task in numerical solution methods of fluid mechanics. Improper mesh designs lead to inefficient, inaccurate and even unstable calculation steps that result in slow, erroneous solutions or even crash in computation. For complex particle shapes that we are dealing with here, proper design of numerical meshes is very crucial.

Since we are dealing with different ice particle shapes here, the mesh designs are different in each case. By mesh deign, we are referring to the specification of computational domain properties. Once these properties are specified, ANSYS Fluent has automatic algorithms to generate the meshes to fit these specified properties. In the following, we use the specification of the computational domains for the columnar ice crystal case [4] as the example to illustrate the computational domain specification.

The computational domain in this case consists of five zones (Fig. 3.3). Division of computational domain is an important strategy in complex flow field situation like this in order to constrain the area of dynamic mesh and maintain the mesh quality. It also saves computer resource by focusing the intensive calculations on the important domain—the one closest to the ice particle—where high grid resolution is necessary

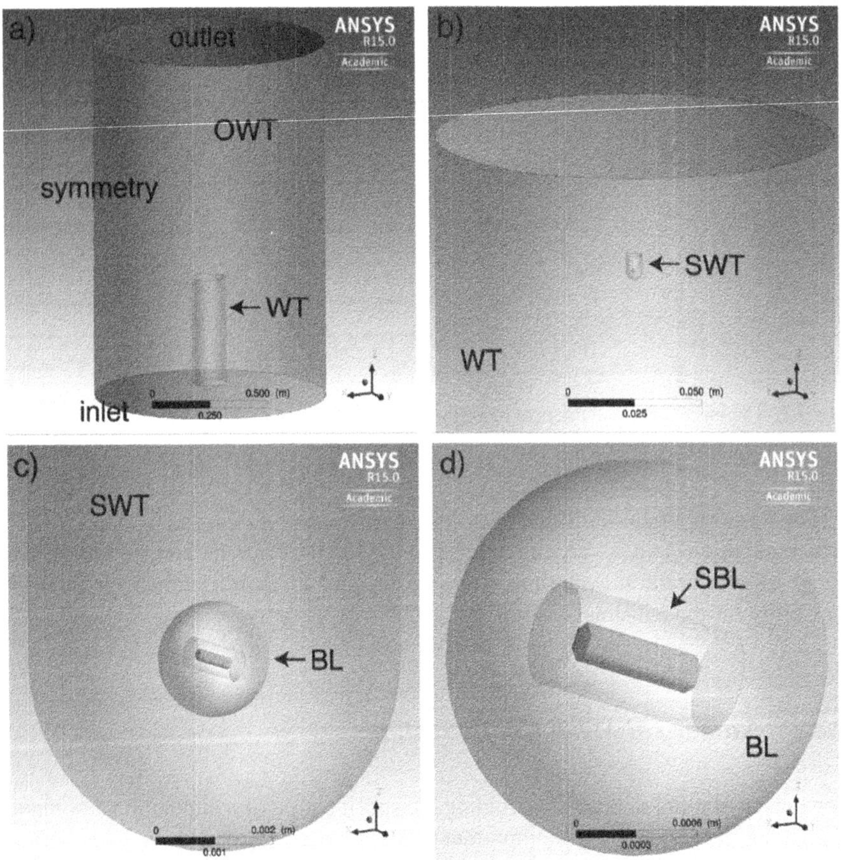

Fig. 3.3 Configuration of computational domain. The cylindrical domain is made up of five fluid zones, outer wind tunnel (OWT), wind tunnel (WT), sub-wind tunnel (SWT), boundary layer zone (BL), and sub-boundary layer zone (SBL), and four boundaries, inlet, outlet, symmetry, and crystal surface, to specify boundary conditions. For details, please see Hashino et al. [4] (adapted from [4], with changes)

to ensure accuracy, and allows coarser resolutions on the outer regions of the flow where changes in flow field are relatively small. In Fluent, the largest computational domain is called the outer wind tunnel (OWT). It has three boundaries, namely, inlet, outlet, and symmetry (or side wall). The mesh inside this zone is stationary to ensure numerical stability near the boundaries. The wind tunnel (WT) is situated inside of OWT, and it allows the mesh to deform as the next interior zone, sub-wind tunnel (SWT), moves. SWT is specified to move horizontally or vertically, according to movement of the ice particle inside, but does not rotate. The mesh in SWT is not affected by the movement of ice particle. On the other hand, the mesh in the boundary layer zone (BL) deforms to accommodate the rotation of the sub boundary layer zone (SBL). SBL passively moves with the particle and inside of SBL the mesh with the inflation does not change. The relative positions of BL, SBL, and ice particle to SWT always stay the same.

The meshes for different ice particle cases may differ somewhat, but the main idea remains the same. This type of design facilitates the computational efficiency by focusing on the high resolution calculations of the flow field near the ice particle while allows relatively coarse grid far away from the ice surface.

Corresponding to the design of the mesh as described above, the boundary conditions (3.3) and (3.4) can be more precisely specified as the following:

$$u_i = 0 \ (i = 1, 2, 3) \ \text{at the ice particle surface,} \tag{3.5}$$

$$u_3 = \text{constant at the inlet of the outmost boundary from the ice particle,} \tag{3.6}$$

$$\frac{\partial u_3}{\partial z} = 0 \ \text{at the outlet of the outmost boundary from the ice particle,} \tag{3.7}$$

$$u_1 = u_2 = 0, \quad \frac{\partial u_i}{\partial x} = \frac{\partial u_i}{\partial y} = 0 \quad (i = 1, 2, 3) \tag{3.8}$$

at the lateral side of outmost boundary from the ice particle.

3.2.3 Specifying the Shapes of Ice Particles

The inner boundary of the present problem is the ice particle surface where the nonslip condition (3.3) applies. In numerical calculations, we need to specify exactly where the surface is, i.e., we need to tell the numerical program what the ice surface coordinates are. In most cases we are dealing with here, the ice surfaces are specified by mathematical expressions given by Wang [13–15] and Wang and Denzer [18] using a technique called successive modification of simple shapes (SMOSS). These expressions involve only analytic functions and very easy to deal with. Specific expressions for certain ice particles will be presented when individual flow fields are described in Chap. 4. Once the equation representing the surface of the particle is chosen, the surface coordinates can be calculated using the ANSYS DesignModeler module.

For some complicated shapes such as snow aggregates, it is nearly impossible to use simple mathematical expressions to represent their shapes. In that case, we manually input their surface coordinates into the program.

3.2.4 Tait–Bryan Angles

As mentioned before, the free fall motion of an ice particle can have six degrees of freedom. The first 3 degrees are associated with the position of its center of gravity (COG) specified by the x, y and z coordinates of the COG. The next 3 degrees of freedom are the orientations of surface specified by the Tait–Bryan angles described in Landau and Lifshitz [9] and Goldstein [3]. Figure 3.4 uses the ice column case as an example to illustrate these angles. We shall call the coordinate system that

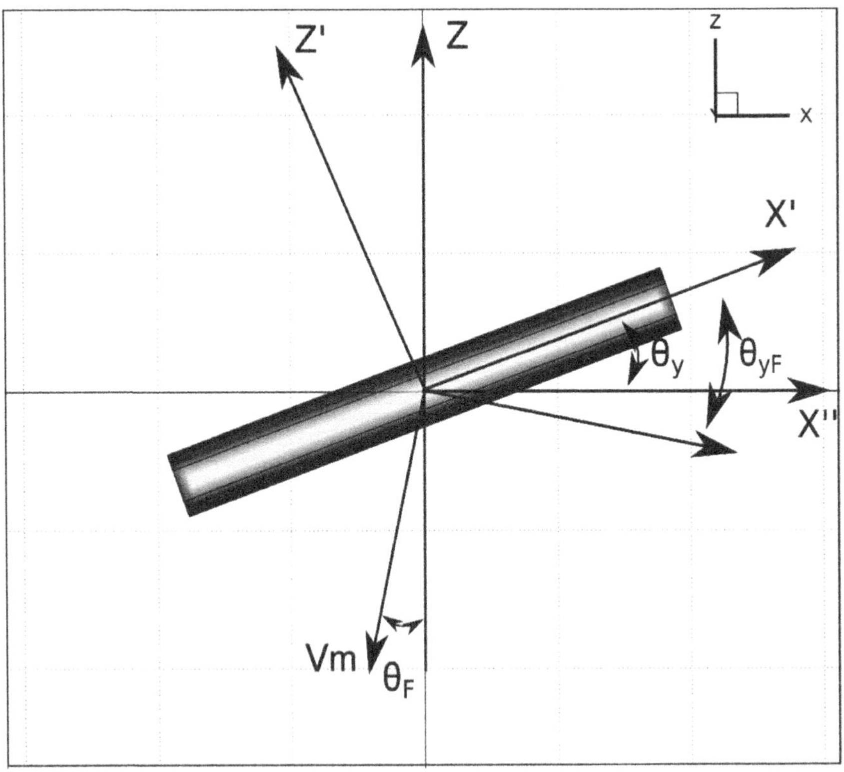

Fig. 3.4 Definition of the Tait–Bryan angles (adapted from [4], with changes)

is stationary with respect to the outer wind tunnel the global coordinate system. Suppose the ice column initially orients with a local coordinate system (x', y', z') that is parallel to the global coordinate system. After the column falls a certain distance, it may rotate so that the column's local coordinate system no longer parallels that of the global coordinate system and each local axis makes a certain angle to the respective global axis. These angles are called the Tait–Bryan angles (i.e., θ_x, θ_y, and θ_z) which are defined as the rotation about the global fixed x, y, and z axes passing through the center of gravity of the column, respectively, as indicated in Fig. 3.4.

Of course, we can make the ice column to tilt in a certain manner initially such that the Tait–Bryan angles are not zero at $t = 0$. An ice column falling in this posture will be subject to a torque that will tend to re-orient the column in a direction different from its original orientation, thus starting an unsteady motion. This is one way to generate asymmetry to initiate unsteady motion in the flow field as will be described below.

3.2.5 *Initial Perturbation*

To induce unsteady motion in the numerical flow field, an initial perturbation—usually a small asymmetry—must be introduced. This asymmetry can be implemented in various ways, for example, Ji and Wang [7] implemented a perturbation by adding an asymmetrical velocity disturbance to the steady state solution of the flow passed a finite length circular cylinder. This induces a periodic vortex shedding in the flow field that persists. In the cases reported in this monograph, when desired, we produce an initial perturbation in the flow field by implementing an initial inclination angle of the particle with respect to an axis. This inclination angle then produces a corresponding perturbation that induces other unsteady flow features. Specific details will be described when individual cases are discussed.

3.2.6 *Instantaneous Velocity, Terminal Velocity and Reynolds Numbers*

In nearly all studies in cloud physics up to the present, especially in the cloud modeling area, cloud and precipitation particles are assumed to fall at their respective terminal velocities at any instant. The terminal velocity $\vec{u}_\infty$ of a falling particle is defined as the velocity at the point when the gravitational force acting on it is balanced exactly by the hydrodynamic drag. This can be described by the equation of motion of a falling particle with mass m_p and velocity $\vec{u}_p$:

$$m_p \frac{d\vec{u}_p}{dt} = m_p \vec{g} + \vec{F}_D = 0 \tag{3.9}$$

where $\vec{F}_D$ is the drag force. The solution of $\vec{u}_p$ in (3.9) is the terminal velocity $\vec{u}_\infty$.

Thus, under the assumption that all particles fall at $\vec{u}_\infty$ all the time, we only need one dimensionless number to characterize the flow field. In this type of fluid mechanical problems, the Reynolds number defined below is the customary choice:

$$N_{\mathrm{Re}} = \frac{du_\infty}{\nu} \tag{3.10}$$

where d is the "size" of the particle and ν is the kinematic viscosity of air. By "size", it is usually taken as the largest dimension of the particle relevant to the flow configuration being considered. For example, d refers to the diameter when the flow past a sphere is of concern. On the other hand, d is the length of a finite cylinder when the cylinder falls with its length oriented horizontally, etc. Sometimes it is difficult to give a precise definition of d when particle shapes become highly asymmetric.

In reality, however, cloud and precipitation particles are not always falling at their terminal velocities. Instead, they can be accelerating or decelerating or even stationary with respect to air at certain instant. This is especially relevant when the particle motion is unsteady with changing fall attitude. When an ice particle performs free fall, its instantaneous velocity can vary significantly from one time step to another due to the changing fall attitude, for example, a column falling with length oriented horizontally would have drag different from when the length is oriented vertically, and thus would have different fall velocities in the two instances. Planar ice crystals are known to perform zigzag motions and their velocities may show periodic variation in both magnitude and direction. Other particles may have their own complicated fall attitudes as described in Chap. 2 before.

In fact, in many cases we have simulated here the particle has never reached true stable terminal velocities. Rather, it may reach a certain quasi-steady state velocity for a while before the velocity gets changed again. In this case, we can define a quasi-steady state Reynolds number as:

$$N_{\mathrm{Re}} = \frac{d(u_\infty - V_t)}{\nu} \tag{3.11}$$

where the relative terminal fall velocity values of V_t are obtained from the simulations that already reach their individual steady-state fall velocity values Vel_z. The quasi-steady state Reynolds numbers are considered as the actual ones that we use to characterize the flow fields in this report.

3.2.7 Computational Strategy

It is useful to make a few remarks here regarding the computational strategy. Our goal is to understand the quasi-steady fall attitude of the ice particles. It will take a long computing time for the ice particle to reach terminal velocity if we begin the computation assuming the particle begins it fall at zero velocity. In addition, the rapid acceleration in the beginning may result in dramatic mesh deformation and increase computational instability. Therefore, we adopt the strategy such that the ice particle has been falling at a reference velocity close to its terminal velocity. This strategy shortens the distance that an ice plate must travel through to reach the quasi-steady state (falling at its terminal velocity), which is the focus of this study. It reduces both the number of time steps and the computational domain size, and accordingly improves the computational efficiency. It also improves the stability of the computations by avoiding the dramatic mesh deformation.

Here we use the case of hexagonal ice plate as an example to illustrate this computational strategy. There are currently no precisely measured terminal velocities of the idealized ice plates we use in the calculations. Thus, we turn to use the terminal velocities of natural plate snowflakes observed by Kajikawa [8] for hexagonal ice plates to obtain the first guess of the reference velocities, and use these velocities as the initial condition for numerical simulations. Based on the reference velocity and the fall motion of the ice plate described by the numerical results, we determine the instantaneous fall velocity of the plate at each time step. From the fall velocity so determined, we assign a new *reference velocity* (the average of the instantaneous velocities) that is close to the true terminal velocity and then rerun the simulation. Such procedure may be repeated several times. We made effort to ensure that the mesh at each time step does not change dramatically from the previous step. This improves the convergence of the transient simulation, reduces the influence of boundary conditions on the ice plates, and hence ensures the quality of the numerical simulation.

For numerical calculations, the resolution of the grid and time can impact the results greatly and some tests are necessary to ensure the accuracy of the results. A grid-independence test was carried for several grid densities and time steps. The baseline configuration was 1.38 million volume cells with a time step of 0.0001 s. A refined grid consisted of 2.56 million cells and a refined time step was 0.00001 s. If the comparison of the baseline results with the refined results indicated no significant difference in the forces exerted on the ice plate, then we choose the baseline configuration to conduct the simulations.

3.3 Ventilation: A Convective Diffusion Problem

A problem that is closely associated with the fall of ice particles is the ventilation effect due to the fall motion. This will be explained in the following. When a hydrometeor is placed in a non-saturated atmospheric environment, phase change of water substance will occur. If the environment is supersaturated, a net migration of water vapor molecules towards the hydrometeor surface will occur, and the hydrometeor will grow in size. On the other hand, if the environment is subsaturated, then evaporation will occur in which there will be a net outward flow of condensed phase water molecules from the hydrometeor surface to become water vapor. It is well known that latent heat will be released during growth and consumed during evaporation.

The migration of water molecules in either towards or away from the hydrometeor surface is a diffusion process. Diffusion will occur even if the hydrometeor is stationary with respect to air as long as the environment is non-saturated. However, the diffusion will be much faster if the hydrometeor is falling. This means that there is an enhancement of the diffusion process due to the motion. This motion-induced enhancement of diffusion is called the ventilation effect, and the factor of enhancement is called the ventilation coefficient.

The magnitude of ventilation coefficient is a function of the vapor density gradient around the hydrometeor. Thus, in order to determine the ventilation coefficient, we first have to determine the vapor density distribution around the hydrometeor and calculate the vapor density gradient.

The vapor density field around a hydrometeor moving at a velocity $\vec{u}$ satisfies the convective diffusion equation for water vapor:

$$\frac{\partial \rho_v}{\partial t} = D_v \nabla^2 \rho_v - \vec{u} \cdot \nabla \rho_v \tag{3.12}$$

where ρ_v is the vapor density and D_v is the diffusivity of water vapor in air. The unsteady form is used because we are dealing with an unsteady transport problem. The specific solution of (3.12) can be obtained once the boundary conditions are specified. The usual boundary conditions are:

$$\rho_v = \rho_{v,s} \text{ at the hydrometeor surface} \tag{3.13}$$

$$\rho_v = \rho_{v,\infty} \text{ far away from the hydrometeor} \tag{3.14}$$

where both $\rho_{v,s}$ and $\rho_{v,\infty}$ are assumed to be constant.

Once ρ_v is determined, the diffusion growth rate of the hydrometeor which is the total flux of water vapor towards the hydrometeor surface can be calculated by

$$\frac{dm}{dt} = -\oint_s (-D_v \nabla \rho_v)_{r=a} \cdot d\vec{S} \tag{3.15}$$

where $d\vec{S}$ represents the increment of drop surface area (see [17, Chap. 9]). The minus sign in front of the surface integral indicates that the "inward" flux is calculated here. Finally, the ventilation coefficient is determined by

$$\overline{f}_v = \frac{(dm/dt)}{(dm/dt)_0} \tag{3.16}$$

where $(dm/dt)_0$ represents the growth rate when the hail is stationary. $\overline{f}_v$ is the mean ventilation coefficient (called ventilation coefficient hereafter) for vapor diffusion.

As mentioned before, there will be release or consumption of latent heat during phase change of water substance whether in diffusion growth or evaporation process. This latent heat will change the temperatures of both the hydrometeor surface and the surrounding air which, in turn, will change the saturation vapor density. Thus the temperature and vapor density fields are coupled in this process. The temperature distribution is also governed by the convective diffusion equation (sometimes called convective conduction equation):

$$\frac{\partial T}{\partial t} = \kappa \nabla^2 T - \vec{u} \cdot \nabla T \tag{3.17}$$

where T is the temperature and κ is the thermal conductivity of air. The usual boundary conditions are:

$$T = T_s \text{ at the hydrometeor surface} \tag{3.18}$$

$$T = T_\infty \text{ far away from the hydrometeor} \tag{3.19}$$

where both T_s and T_∞ are assumed to be constant. Clearly, the boundary conditions (3.18)–(3.19) are the same as (3.13)–(3.14) mathematically.

Thus, we see that the heat diffusion phenomenon is entirely analogous to water vapor diffusion which means that the heat transfer rate (heating or cooling rate) will also be enhanced by a ventilation coefficient relevant to heat diffusion $\overline{f}_h$ due to the motion of the hydrometeor. It is usually assumed that

$$\overline{f}_v = \overline{f}_h \tag{3.20}$$

(see [11]).

3.4 Terminal Velocity

We have mentioned the terminal velocity previously. In this section, we shall examine it more quantitatively and develop further on it. Again, a cloud or precipitation particle reaches the terminal velocity when the gravitational force is balanced by the drag force, that is,

$$m\frac{d\vec{u}}{dt} = \vec{F}_g + \vec{F}_D = 0 \tag{3.21}$$

where m is the particle mass, $\vec{F}_g$ the buoyancy-corrected gravitational force and $\vec{F}_D$ the drag force of air acting on the particle. Expanding explicitly, (3.21) becomes

$$m\frac{d\vec{u}}{dt} = m\vec{g}\left(\frac{\rho_w - \rho_a}{\rho_w}\right) + \vec{F}_D = 0 \tag{3.22}$$

where ρ_w and ρ_a are the density of liquid water and air, respectively. (3.22) becomes

$$m\vec{g}\left(\frac{\rho_w - \rho_a}{\rho_w}\right) = -\vec{F}_D \tag{3.23}$$

For a small cloud drop, we can use the Stokes drag $\vec{F}_D = 6\pi\eta a u_\infty$ where η is the dynamic viscosity of air and a the radius of the drop (see [17, p. 191]). Thus we have

$$mg\left(\frac{\rho_w - \rho_a}{\rho_w}\right) = 6\pi\eta a u_\infty \tag{3.24}$$

and

$$u_\infty = \frac{mg\left(\frac{\rho_w-\rho_a}{\rho_w}\right)}{6\pi\eta a} = \frac{2a^2 g(\rho_w - \rho_a)}{9\eta} \approx \frac{2a^2 g\rho_w}{9\eta} \tag{3.25}$$

(3.25) can be used to estimate the terminal velocity of small particles.

For larger spherical particles, the drag force is expressed as

$$\vec{F}_D = 6\pi\eta a u_\infty\left(\frac{C_D N_{\mathrm{Re}}}{24}\right) \tag{3.26}$$

and hence (3.24) becomes

$$mg\left(\frac{\rho_w - \rho_a}{\rho_w}\right) = 6\pi\eta a u_\infty\left(\frac{C_D N_{\mathrm{Re}}}{24}\right) \tag{3.27}$$

And

$$u_\infty = \frac{mg\left(\frac{\rho_w-\rho_a}{\rho_w}\right)}{6\pi\eta a\left(\frac{C_D N_{\mathrm{Re}}}{24}\right)} \tag{3.28}$$

Thus the terminal velocity is a function of the drag coefficient C_D.

The drag coefficient C_D of non-spherical particles is complicated as it is not only a function of fall speed and particle shape but also a function of fall attitude. For example, an ice plate falling with basal face pointing downward will have a drag coefficient different from when the plate falls with basal face oriented obliquely. Consequently, the terminal velocity of the plate will change when its orientation changes.

It is generally difficult to obtain C_D via analytical solutions for large highly nonspherical particles such as the ice particles we are dealing with in this monograph. We use entirely numerical methods to determine the flow fields and C_D for ice particles falling freely in air and the results will be reported in Chap. 4.

References

1. Cheng KY, Wang PK, Hashino T (2015) A numerical study on the attitudes and aerodynamics of freely falling hexagonal ice plates. J Atmos Sci 72:3685–3698
2. Freitas CJ, Street RL, Findikakis AN, Koseff JR (1985) Numerical simulation of three-dimensional flow in a cavity. Int J Numer Methods Fluids 5:561–575
3. Goldstein H (1980) Classical mechanics, 2nd edn. Addison-Wesley, 672 pp
4. Hashino T, Cheng K-Y, Chueh C-C, Wang PK (2016) Numerical study of motion and stability of falling columnar crystals. J Atmos Sci 73:1923–1942
5. Issa RI (1986) Solution of the implicitly discretised fluid flow equations by operator-splitting. J Comput Phys 62:40–65
6. Ji W, Wang PK (1990) Numerical simulation of three-dimensional unsteady viscous flow past fixed hexagonal ice crystals in the air—preliminary results. Atmos Res 25:539–557. https://doi.org/10.1016/0169-8095(90)90037-D
7. Ji W, Wang PK (1991) Numerical simulation of three-dimensional unsteady viscous flow past finite cylinders in an unbounded fluid at low intermediate Reynolds numbers. Theor Comput Fluid Dyn 3:43–59
8. Kajikawa M (1972) Measurement of falling velocity of individual snow crystals. J Meteor Soc Jpn Ser II 50:577–584
9. Landau LD, Lifshitz EM (1969) Mechanics, 2nd edn. Pergamon Press, 165 pp
10. Pitter RL, Pruppacher HR (1974) A Numerical investigation of collision efficiencies of simple ice plates colliding with supercooled water drops. J Atmos Sci 31:551–559
11. Pruppacher HR, Klett JD (1997) Microphysics of clouds and precipitation, 2nd edn. D. Reidel, New York, 954 pp
12. Schlamp RJ, Pruppacher HR, Hamielec AE (1975) A numerical investigation of the efficiency with which simple columnar ice crystals collide with supercooled water drops. J Atmos Sci 32:2330–2337
13. Wang PK (1982) Mathematical description of the shape of conical hydrometeors. J Atmos Sci 39:2615–2622
14. Wang PK (1997) Characterization of ice crystals in clouds by simple mathematical expressions based on successive modification of simple shapes. J Atmos Sci 54:2035–2041
15. Wang PK (1999) Three-dimensional representations of hexagonal ice crystals and hailstone particles of elliptical cross sections. J Atmos Sci 56:1089–1093
16. Wang PK (2002) Ice microdynamics. Academic Press, 273 pp
17. Wang PK (2013) Physics and dynamics of clouds and precipitation. Cambridge University Press, 467 pp
18. Wang PK, Denzer SM (1983) Mathematical description of the shape of plane hexagonal snow crystals. J Atmos Sci 40:1024–1028
19. Wang PK, Ji W (1997) Numerical simulation of three-dimensional unsteady flow past ice crystals. J Atmos Sci 54:2261–2274

Chapter 4
Flow Fields and Fall Attitudes of Ice Hydrometeors

In this chapter, we will present the numerical simulation results of the flow fields around freely falling ice hydrometeors. The physics and numerical methods of these simulations have been described in Chap. 3. We will report the flow fields for different kinds of ice particles separately. The ice particle types whose fall behaviors to be reported include hexagonal plates, hexagonal columns, stellar crystals, broad branch crystals, sector plates, simple dendrites, aggregates of dendrites, conical graupel and spherical and lobed hailstones. While it is impossible to cover a comprehensive range of particle shapes which the number must be in the thousands, we believe that the results presented here represent the widest coverage of theoretical flow fields around falling large ice particles and they should be adequately representative for general parameterization purpose for inclusion in cloud models.

Note that the atmospheric environments in which these ice particles fall may not be the same as each kind of ice hydrometeors has its favorable environment of existence different from other ice particle types. However, what we emphasize here is the hydrodynamic behaviors of the fall of these particles. The effect of specific atmospheric environment, while quantitatively important, does not impact significantly the qualitative understanding of the hydrodynamic behaviors and the latter is the focus of this monograph.

4.1 Flow Fields Around Freely Falling Hexagonal Ice Plates

Hexagonal plates are one of the most common habits of ice crystals and also one of the simplest. In this study, we use a regular hexagonal plate (P1a of Mogono-Lee classification, see Fig. 4.1) of uniform thickness to approximate the shape of the ice plate. The density of the ice plate is assumed to be 0.91668 kg m^{-3}, i.e., the general solid ice density. Hollow ice plates would have lower density but they are not investigated here. The ice plate is assumed to fall in the atmospheric environment of air density $\rho_a = 1.3224$ kg m^{-3} and kinematic viscosity $\nu = 1.261515 \times 10^{-5}$ m^{2}s^{-1}

© Springer Nature Singapore Pte Ltd. 2021

P. K. Wang, *Motions of Ice Hydrometeors in the Atmosphere*, Atmosphere, Earth, Ocean & Space, https://doi.org/10.1007/978-981-33-4431-0_4

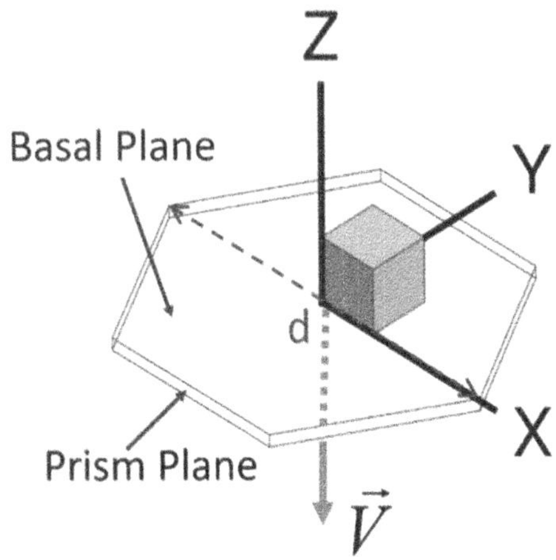

Fig. 4.1 Geometrical configuration of the problem

(approximating the winter condition with 1000 hPa and −10 °C). We also assume that the ice plate initially falls vertically with the basal plane oriented horizontally, i.e., parallel to the x–y plane of the global coordinate system, as shown in Fig. 4.1.

4.1.1 Dimensions of Ice Plates

In the following, we summarize the flow fields of falling ice plates with diameter 1, 2, 3, 4, 5, 7, and 10 mm. The "diameter" used here refers to the diameter of circumcircle of the ice plate, as indicated in Fig. 4.1. The aspect ratio (thickness to diameter) of natural ice particles is not arbitrary but follows certain empirical relations as studied by Auer and Veal [2]. For ice plates, the thickness is determined by the rounded aspect ratio (thickness to diameter) based on the empirical formula proposed by Auer and Veal [2]:

$$h = 2.020d^{0.449}, \tag{4.1}$$

where h is the thickness and d is the diameter. Both units are in μm.

4.1.2 Terminal Velocities of Falling Ice Plates of 1–10 mm Diameter

In contrast to all previous studies of steady state flows or unsteady but fixed ice particle orientation flows (e.g., [54]) where there is a definite terminal speed in each case, the ice particles studied in this monograph in general and the ice plates discussed here in particular, the fall speeds do not remain constant during simulations but rather keep fluctuating up and down throughout the fall process, only that the fluctuation is bounded within a certain range. The 2 mm plate is the only exception that its fall velocity shows a slight increasing trend. The exact reason for the trend is still unclear but is likely due to the large inclination during its fall. Larger inclination results in smaller projected area of the ice plate normal to the vertical wind that reduces the drag. Consequently, the plate gains larger downward acceleration.

Due to the huge computing resource required, we simulate the fall of each ice plate for 0.5 s for each case. We average the fall velocities at the last 3000 time steps to represent the terminal velocity when the fall attitude of the plate becomes quasi-periodic. The variability of the fall velocity is less than 4% for each case, including the 2 mm case. Figure 4.2 shows the calculated terminal velocities as a function of the plate diameter. Also plotted in this chart are the terminal velocities estimated from the observational data given by Locatelli and Hobbs [29] and Heymsfield and Kajikawa [15] for ice plates. Clearly, the two observational studies show a similar trend as the present study. Both theoretical results and observational data show that the terminal velocity of hexagonal ice plate increases with diameter but the increasing rate decreases. If the trend continues, the terminal velocity curve would eventually reach a plateau. If this happens, it will be due to the increase in drag for larger plates that compensates for larger gravitational force, just like the terminal velocity of large

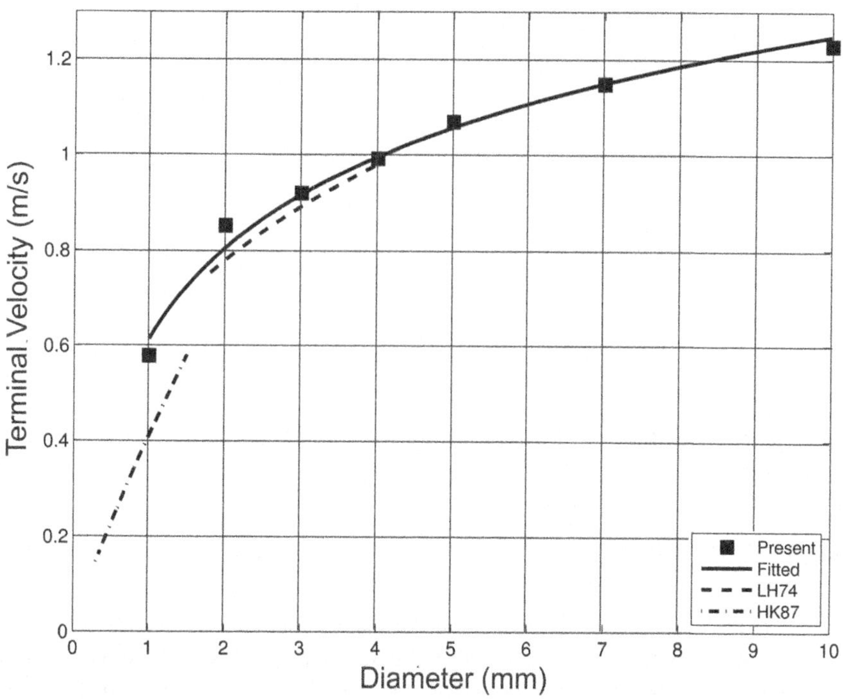

Fig. 4.2 Terminal velocities of ice plates (adapted from [7], with changes)

Table 4.1 List of diameters, aspect ratios, terminal velocities, and Reynolds numbers of the ice plates investigated here (adapted from [7], with changes)

Diameter (mm)	Aspect ratio	Terminal velocity (m s^{-1})	Reynolds number
1	0.0449	0.578	46
2	0.0307	0.851	135
3	0.0245	0.917	218
4	0.0209	0.991	314
5	0.0185	1.069	424
7	0.0154	1.148	637
10	0.0126	1.228	974

water drops reaching a plateau as shown in Wang and Pruppacher [56]. But currently we have not yet done calculations for $d > 1$ cm.

It is seen that the theoretical terminal velocities are slightly higher than observational ones. This is likely due to the difference in shape as the natural ice plates often differ from an exact hexagon. Natural ice plates are often not exact hexagons but may have branches grown on the vertices. In that case, the volume (and the mass, assuming same density) of a natural ice plate will be smaller than that of an exact hexagon if both has the same thickness. Hence it is probably not surprising that the theoretical exact hexagonal plates have slightly higher terminal velocities than that of natural ice crystals.

The calculated terminal velocities (u_∞) can be fitted by the exponential equation:

$$u_\infty = 0.2755 \ln(9.271d), \tag{4.2}$$

where u_∞ has the unit of m s^{-1}.

The moment of inertia of a particle is sometimes useful for the discussion of motion attitude and is calculated. The dimensionless moment of inertia is defined as

$$I^* = \frac{I_a}{\rho_a d^5}, \tag{4.3}$$

where I_a is the moment of inertia about the a-axis of the ice plates (Table 4.1).

4.1.3 *Fall Attitudes*

To see if the calculated fall behavior of the plates simulates reasonably well with reality, we compare the simulated fall attitudes with the observed. Kajikawa [25] studied falling natural planar snow crystals and found that hexagonal ice plates (P1a) exhibit a stable fall motion (falling with the c-axis vertical) for $N_{\mathrm{Re}} < 47$. When $N_{\mathrm{Re}} \geq 47$, the crystals began to perform unsteady fall motions including oscillation, horizontal translation, bell swing, and spiral motion. List and Schemenauer [28] used planar crystal models to experimentally study the fall behavior of hexagonal plates (P1a) and found that the plate started to oscillate at $N_{Re} > 100$. The numerical results obtained in this study are consistent with the both observations of Kajikawa [25] and List and Schemenauer [28]. The simulated motion of 1 mm ice plate ($N_{Re} = 46$) exhibits a steady fall motion—it falls vertically with almost no horizontal displacement and oscillation. The rest cases (for $d = 2$–10 mm, $135 \leq N_{Re} \leq 973$) exhibit unsteady fall attitudes with noticeable horizontal movements and evident oscillations.

As an example, Fig. 4.3 shows the consecutive fall attitudes of the simulated 5 mm ice plate during the period from 0.2 to 0.5 s with an interval of 0.025 s. The ice plate

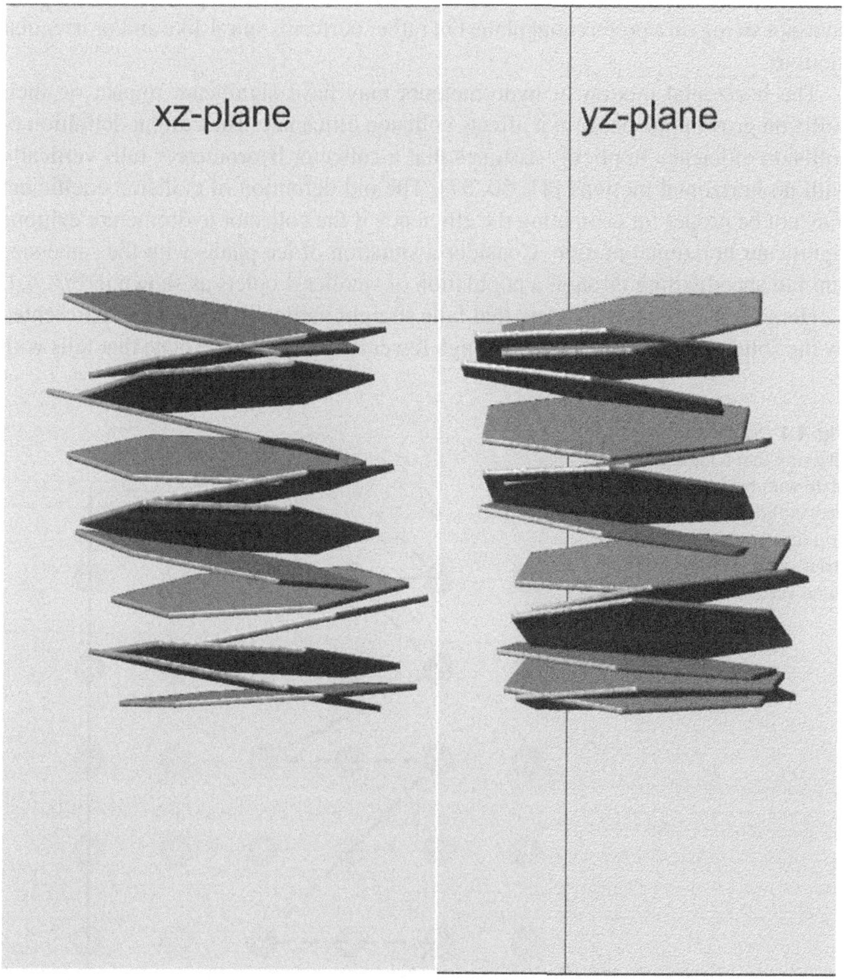

Fig. 4.3 Succesive position and orientation of the falling plate in xz- and yz-plane for t = 0.1–0.4 s, interval 0.025 s

oscillates about an a-axis roughly normal to the x–z plane and also exhibits a zigzag swing in the x–z plane. The horizontal displacement due to the swing in this case is about 0.2 d. More will be said about this displacement.

There is a coordinated action between the oscillation and the swing motion. When the ice plate tilts to the right (left), the ice plate tends to move rightward (leftward). This behavior occurs to all unsteady cases studied here, and is consistent with the experimental studies of Willmarth et al. [58] and Stringham et al. [46] who experimentally investigated the motions of freely falling circular disks and found that the most common unsteady motion for a falling disk is an oscillation accompanied by a horizontal motion.

For the smaller size of 2 and 3 mm cases, the ice plate swings back and forth quasi-steadily on a preferential vertical plane. For larger ice plates, the ice plate does not swing on a preferential plane but rather performs spiral-like and/or irregular motions.

The horizontal motion of hydrometeors may have significant impact on their collision growth rate because it affects collision efficiency. The current definition of collision efficiency implicitly assumes that a collector hydrometeor falls vertically with no horizontal motions [41, 50, 57]. The old definition of collision coefficient may not be proper for estimating the efficiency if the collector hydrometeor exhibits significant horizontal motion. Consider a situation of ice plates with the same size and fall speed falling through a population of smaller droplets as shown in Fig. 4.4.

Here we see that the ice plate that falls straight vertically (trajectory represented by the solid arrow) would sweep through fewer droplets than the plate that falls with

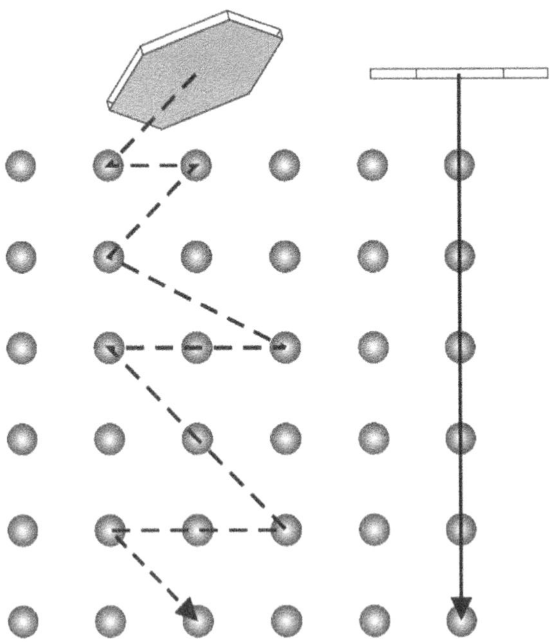

Fig. 4.4 Schematic chart showing that a falling ice plate with horizontal motion may collide with more droplets than a plate with strict vertical fall over the same vertical distance

Table 4.2 Sumary of ice plate fall behavior (after [7] with changes)

Diameter (mm)	Dimensionless displacement l^*	Vibration frequency about x and y-axis (Hz)	Vibration frequency about z axis (Hz)	Characteristic ϕ (°)	Characteristic θ (°)	Characteristic ψ (°)
1	0.042	29.30	N/A	0.01	0.00	N/A
2	0.414	18.31	36.62	18.01	3.77	0.62
3	0.217	14.65	29.30	6.63	1.92	0.08
4	0.084	12.21	24.41	2.87	2.05	0.05
5	0.337	10.99	21.97	6.64	10.81	0.41
7	0.301	8.54	15.87	8.56	6.64	0.66
10	0.101	6.10	12.21	2.15	3.23	0.14

horizontal displacement (trajectory represented by the dashed arrow). Although this figure is grossly simplified, it does show that ice plates with horizontal motions, compared with those with pure vertical motion, will move longer distance and hence collide with the droplet at a different efficiency. In addition, the oscillation of the ice plate affects the collision efficiency by changing the flow field around it and its cross-sectional area normal to the vertical direction. Indeed, there is a need to study the impact of horizontal motions and orientations on collision efficiency for all larger ice hydrometeors in order to assess their collision grow more properly.

To characterize this horizontal movement, we define a dimensionless horizontal displacement $l^* = l/d$ where l is the characteristic length of the swing, spiral and irregular motions to characterize the extent of this horizontal motion. Table 4.2 shows that l^* for ice plates ranges from 0.084 to 0.414. There seems no systematic relationship between l^* and d for the ice plates studied here.

Hexagonal plates are fairly common in atmospheric clouds and they may cause optical phenomena such as halo and specular reflection under suitable conditions, and understanding these conditions in relation to the ice plates' physical properties may help atmospheric remote sensing techniques (e.g., [18, 31, 39]). The refraction of light by an ice plate is influenced by its orientation which, in turn, is influenced by the fall attitude of ice crystals.

For ice plates studied here, we can use the Tait–Bryan angles (ϕ, θ, and ψ) defined in Sect. 3.2.4 to represent the orientation of an ice plate with respect to its initial orientation. As a reminder, ϕ, θ, and ψ are defined as the rotation angle about the x, y, and z axes of the ice plate, respectively. An example of time–varying Tait–Bryan angles is shown in Fig. 4.5.

Here we see that all three rotation angles show periodic behavior. The ϕ and θ oscillate at about the same frequency initially but θ oscillates with larger amplitudes (note the different scales of the angle). Later the frequencies become different. Some of the cases take their own preferential a-axis to oscillate even though the moments of inertia about any a-axes are all the same. One possible explanation is that the

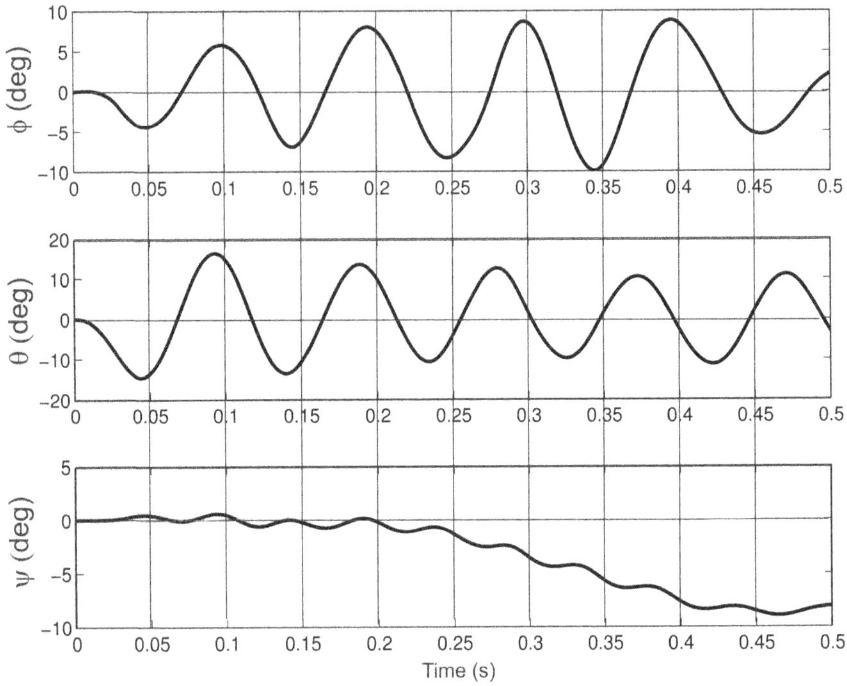

Fig. 4.5 Tait–Bryan anles as a function of time for a falling hexagonal ice plate of 5 mm diameter (adapted from [7], with changes)

hexagonal ice plates used in numerical simulations are not perfectly symmetric as the meshing technique does not produce a perfectly symmetrical hexagon. Even a slight asymmetry may cause uneven mass distribution and one of the *a*-axes becomes the preferred axis. Another possible explanation is that the flow field around an ice plate is asymmetrical in nature. This asymmetry may cause the ice plate to oscillate along a preferential *a*-axis. Finally, ψ oscillates at higher frequency but with much smaller amplitudes than do ϕ and θ. The behavior described in this paragraph is valid for all cases examined in this study.

Table 4.2 shows the characteristic oscillation frequencies and corresponding amplitudes (angles) for the ice plates studied here. We note that the vibration frequencies about the x and y axes are the same, which is likely due to that the moments of inertia about any *a*-axis of a regular hexagonal plate are the same. The oscillation frequency decreases as the diameter of the ice plate increases (Fig. 4.6). The relationship between the oscillation frequency (about *a*-axis) f and diameter d can be fitted by the following empirical formula:

$$f = 30.22d^{-0.5978} - 1.079 \tag{4.4}$$

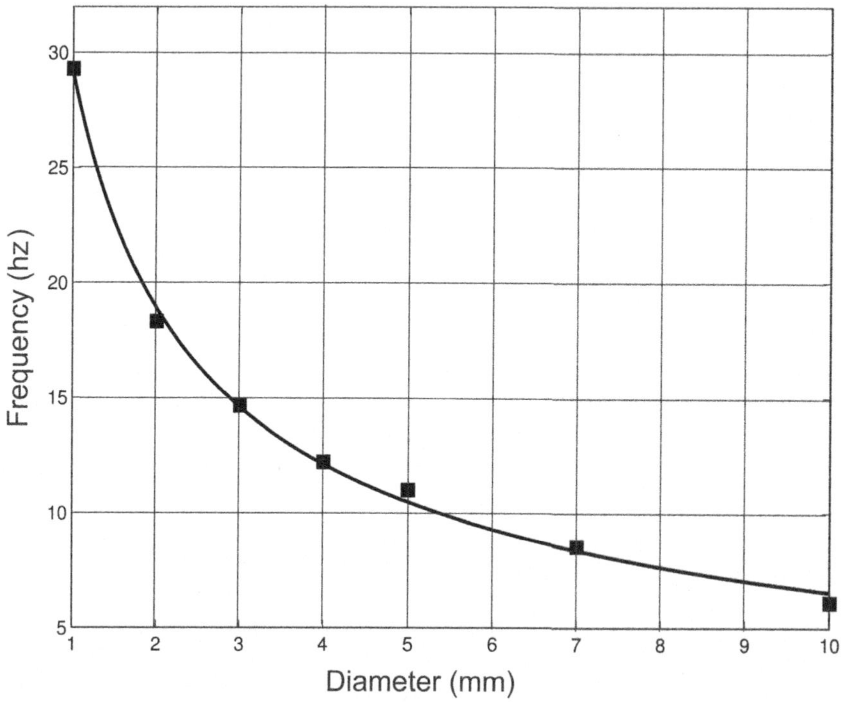

Fig. 4.6 Oscillation frequency of ice plates as a function of diameter (adapted from [7], with changes)

Comparing the vibration frequency about the z-axis to that about the x and y axes reveals that the vibration frequency about the z axis is about twice as that about the x and y axes.

The fact that the characteristic angles ϕ or θ are much larger than ψ indicates that these ice plates prefer to oscillate about the a-axis rather than the c-axis. The characteristic angles about the a-axis are from 1.92° to 18.01°, but similar to the dimensionless horizontal displacement l^*, they show no unique relationship with the dimension of the ice plates. Neither do they seem to be a function of the moment of inertia. However, the maximum characteristic angles of ϕ or θ decrease with the decreasing dimensionless horizontal displacement. The plate orientation will certainly impact its horizontal displacement. Given the same external force normal to the plate, an ice plate with higher inclination will experience a larger horizontal force component and hence executes greater horizontal displacement.

The relation of non-dimensional horizontal displacement versus angle can be fitted by

$$l^* = -4.131\Theta^{-0.04938} + 4.001, \tag{4.5}$$

where Θ is the maximum of the characteristic angle ϕ and θ.

The characteristic value of ψ is determined from the high frequency signal. For some cases, low frequency signals seem to be the main feature of ψ, e.g., the lowest panel in Fig. 4.5. Unfortunately, the simulation time of the present study is not sufficiently long to decide the role of the low frequency signal on ψ.

4.1.4 Characteristics of Flow Around Falling Ice Plates

Figure 4.7 shows a randomly selected frame of the flow field around a falling hexagonal ice plate of $d = 1$ mm. It shows the velocity vectors and the pressure deviation on an y–z plane crossing the center of the ice plate at this time. As mentioned before, previous observations show that the fall is steady for this case. The numerical results also show that the flow field quickly becomes steady state. Note that the vectors are the projection of the 3-D vectors on this plane as the real vectors are 3-dimensional

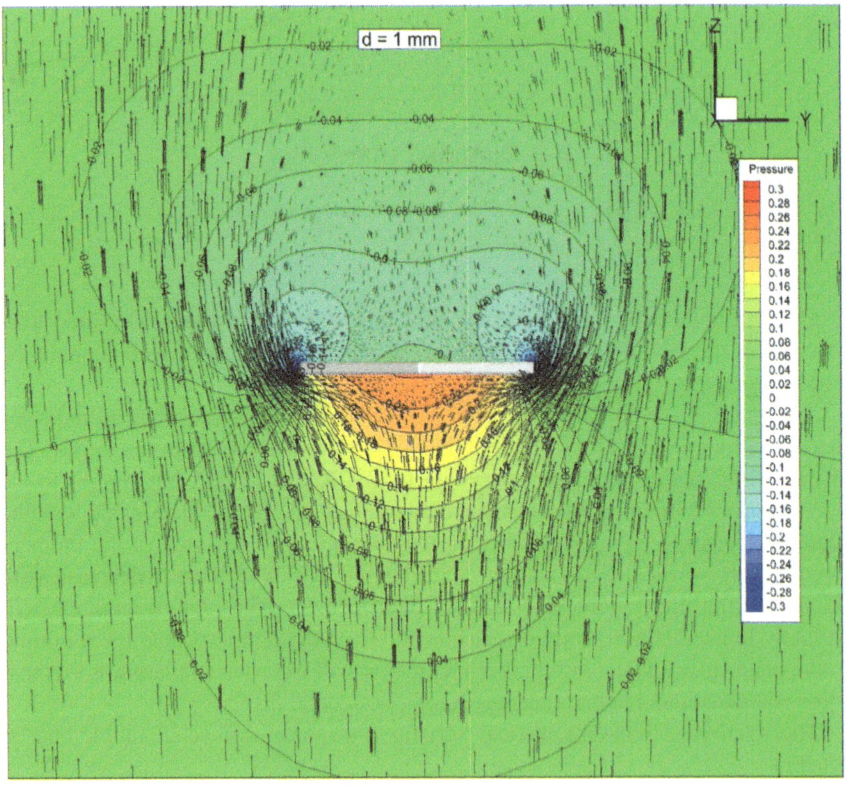

Fig. 4.7 The pressure distribution and velocity vectors in the yz plane of a falling hexagomnal ice plate of 1 mm diameter (after from Cheng et al. with changes)

and have components normal to the y–z plane. The uneven vector distribution is due to uneven mesh setup and has nothing to do with the velocity magnitude.

The pressure distribution shows that the positive pressure deviations occur in the upstream region with the pressure maximum occurs right below (the upstream side) the ice plate. The negative pressure deviations generally occur in the downstream wake region with the absolute pressure minimum occurs at the upper edge of the ice plate. Such pressure distribution is generally similar to that in the studies of Pitter et al. [38], Ji and Wang [21, 20], Wang and Ji [54] and Hashino et al. [14]. However, one notable difference between the pressure distribution of Pitter et al. [38] and that of all other later studies mentioned above is that the outermost edge of the underside of the ice plate in all later studies is always associated with negative pressure deviation (see Fig. 4.8), whereas it is just the opposite in Pitter et al. [38]. This is because Pitter et al. [38] approximated the hexagonal plate by a thin oblate spheroid that has only one sharp edge and the underside surface is associated with positive pressure deviation. The ice plates in all later studies are assumed to be hexagonal plates of finite thickness and hence each has two sharp edges. This fact may be of importance

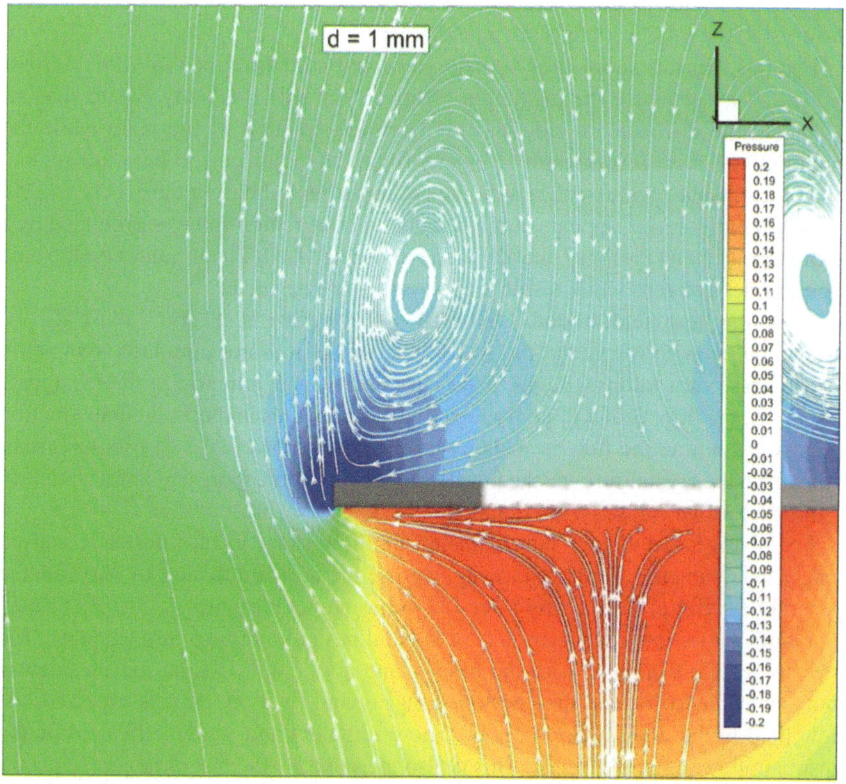

Fig. 4.8 An enlarged view of the pressure distribution around the corner of a falling ice plate of 1 mm diameter. White curves are streamtraces (after [7] with changes)

since the rim of the ice plate is where a supercooled droplet is most likely to hit the plate and form rime, and the difference in pressure distribution will likely cause a difference in collision efficiency that alters the riming rate and the location where the drop will land on the plate. This rim pressure configuration occurs in all the plate cases in the present study.

To see the flow field and the pressure distribution in more detail, a few streamtraces are plotted Fig. 4.8. From this figure, one can see that if a small particle is initially located near the center in the underside of the plate and travels along one of the streamtraces, it will very likely land at the underside edge of the plate instead of hitting the plate at the center. This indeed happens in clouds as the initially riming of small ice crystals occurs preferentially along the underside edge of an ice plate crystal (e.g., see Fig. 10.9 of [57]).

Figure 4.8 also shows that a pair of standing eddies occur in the wake region which is a standard feature of the steady flow field of some falling hydrometeors which is sometimes called the *recirculation bubble*. Inside the bubble, the velocities are smaller as we move inward and become downward as we move near the center. The flow velocities close to both the upper and lower basal planes of the ice plate are very small due to the obstruction of the crystal. On the side and far away from the crystal, the flow velocities are much larger.

Figure 4.9 further shows the z-velocity (w) distribution in the central vertical cross-section. It is seen here the maximum positive (upward) z-velocity (white shaded) occurs in the downstream side and off from the position of the plate while the maximum negative (downward) z-velocity (blue shaded) occurs in the wake region. The blue-shaded region is all within the recirculation bubble. In the 3-D sense, the velocity configuration appears as a low velocity cone region (blue) surrounded by a high velocity bowl (yellow) outside as shown in Fig. 4.10. The w values on the blue cone surface are zero while inside it all w values are negative.

Since the flow field around a falling 1 mm ice plate is steady, it must also be axisymmetric with respect to the z-axis. Of course, it is not circularly symmetric like the case of a falling circular plate but rather it is symmetric in the hexagonal sense, i.e., the flow field will go through 6 identical cycles in the azimuthal direction when we examine the flow near the crystal surface. However, as we move further away from the crystal surface, the flow field would become more and more circularly symmetric as we can see from Fig. 4.10.

For plates larger than 1 mm diameter, the flow starts to turn unsteady and no longer axisymmetric. In the following, we use the $d = 5$ mm case to illustrate the general behavior of the unsteady free fall of hexagonal ice plates.

Figure 4.11 shows two snapshots of streamlines around the 5 mm ice plate. The plate is released initially with the basal plane horizontal. It soon makes sideway movements and tilts, and then performs zigzag and rotation, etc. As expected, the flow field in the downstream becomes more turbulent and asymmetrical while the upstream flow remains laminar. This is also the same for the rest of larger ice plates.

The eddy shedding appears in the downstream of the eddy as an important feature of unsteady flow. In the two snapshots, the ice plate exhibits different orientations and the flows inside the eddy region are different, however, the sizes, as well as

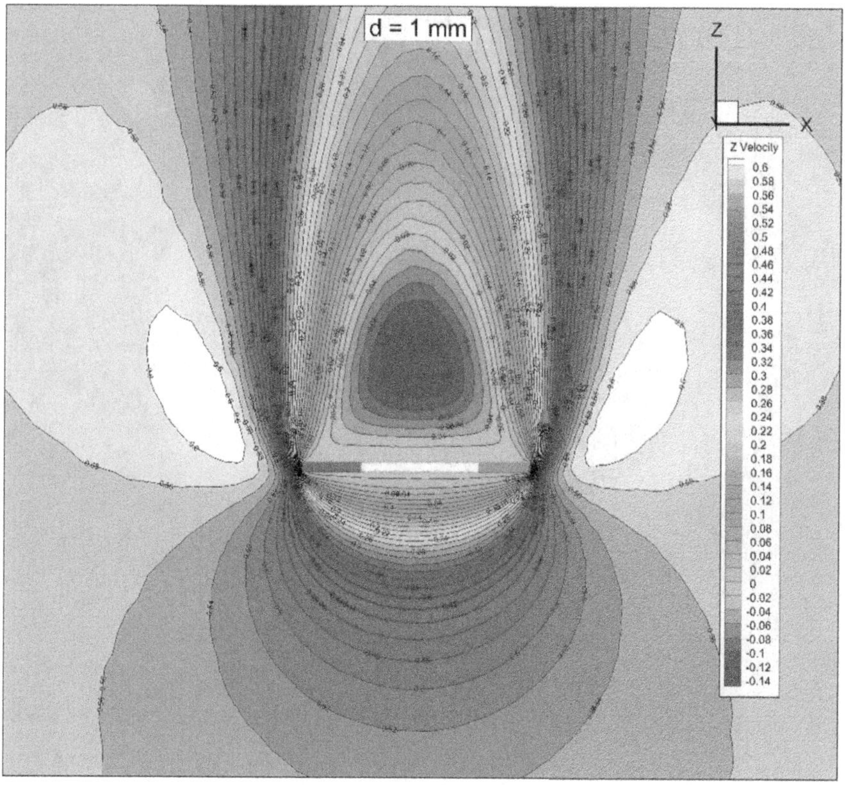

Fig. 4.9 The z-velocity distribution around a falling ice plate of 1 mm diameter

shapes, of the two eddies are about the same. Note that the flow in the upstream no longer remains steady, which is opposite to previous studies that did not allow the hydrometeors to move and rotate freely inside the computational domain (e.g., [14, 27]). The location of the separation of the upstream flow starts to move with the orientation of the ice plate. Such flow pattern, compared to steady upstream cases reported in previous studies, may result in different collision and diffusion growth for an ice plate.

Whilst the upstream flow is unsteady, it is nevertheless laminar and there is no trace of turbulence there.

Figure 4.12 shows the pressure distributions on the x–z plane crossing the center of the ice plate at the same two time steps corresponding to Fig. 4.11. Note that the plate is allowed to move freely in all three axial directions so that the ice plate is not always falling along the central x–z plane, but rather it can move in and out of that plane. Thus, while the pressure field in Fig. 4.12 corresponds to that on the fixed central x–z plane, it does **not** correspond to the field along the vertical plane going through the center of the crystal.

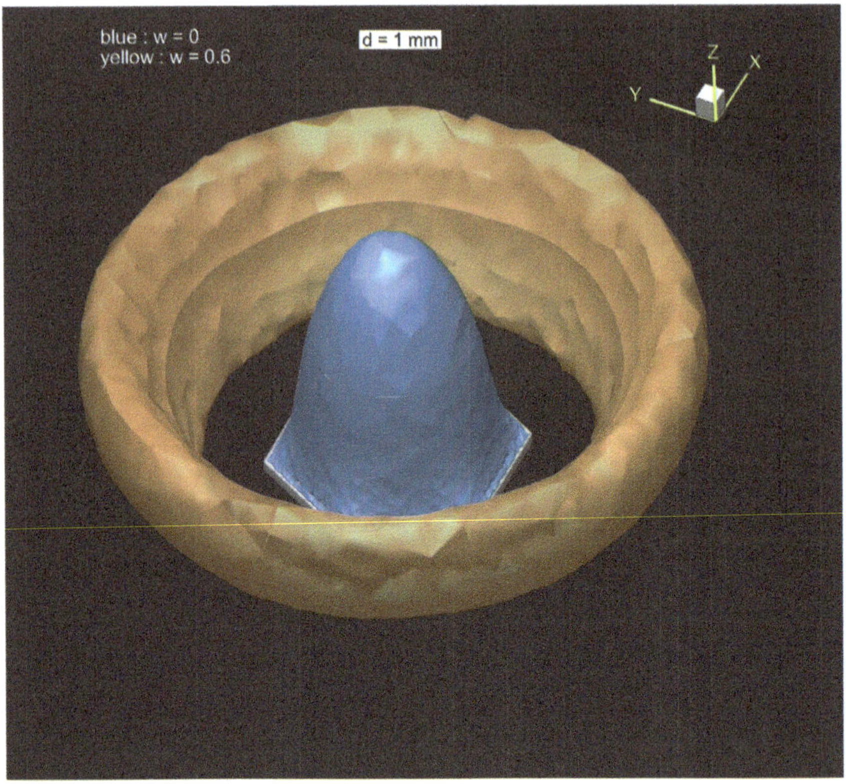

Fig. 4.10 The 3-dimensional distribution of z-velocity around a falling hexagonal ice plate. Negative (downward) velocity region is enveloped by the blue contour surface. Velocities are in cm/s

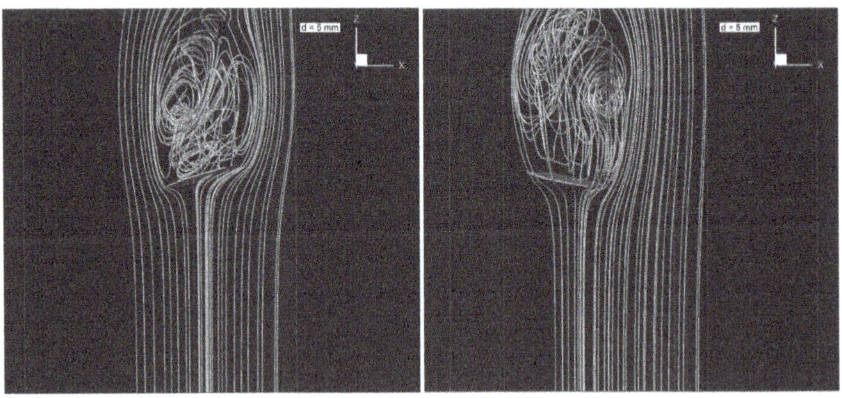

Fig. 4.11 Streamtraces around a faling hexagonal ice plate of 5 mm diameter at two randomly selected time steps showing the turbulent wake (after [7] with changes)

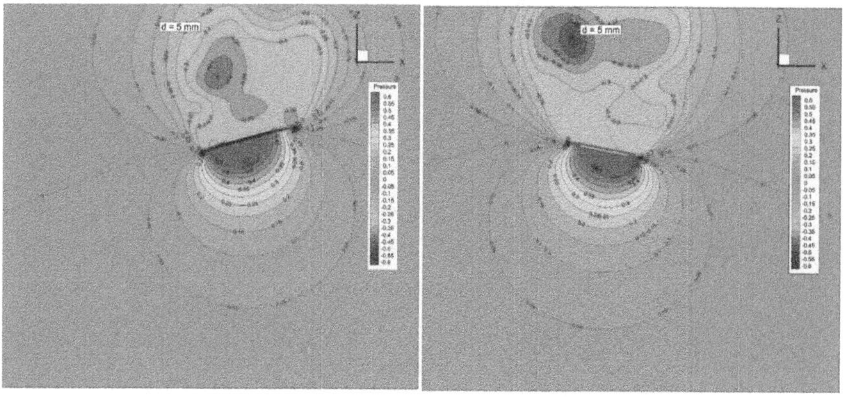

Fig. 4.12 The yz plane pressure distribution around a falling hexagonal ice plate at time step corresponding to that in Fig. 4.11 (after [7] with changes)

Like the $d = 1$ mm case in Fig. 4.7, the fall is always associated with high pressure in the upstream and low pressure in the downstream. But unlike the 1 mm case, the minimum pressure in the wake is not attached to the crystal surface but rather it moves further downstream. This is the well-known **eddy shedding** and the low pressure centers swim away from the surface. The pressure minimum changes its location and magnitude near-periodically, forming the quasi-periodic eddy shedding pattern. However, the eddy shedding frequency seems not to directly relate to the oscillation frequency. Exactly how much impact the oscillation has on the shedding frequency remains to be studied in the future.

Unlike the steady case, the location of the pressure **maximum** does not remain stationary with respect to the ice plate but varies with the plate orientation. When the ice plate tilts to the right (left), the pressure maximum tends to move rightward (leftward). Such asymmetrical pressure distribution may explain the periodic rotations and horizontal motions. When the ice plate tilts to the right (as indicated in the left panel of Fig. 4.12), the asymmetrical pressure below the ice plate will generate a **torque** pointing roughly out of the y–z plane, twist the ice plate counterclockwise, and push the plate to tilt back to the left. The same pressure distribution will also generate a leftward horizontal force, and push the plate in the left direction. In other words, once an ice plate tilts, the uneven pressure distribution below such ice plate will tend to twist and move it toward the opposite inclination and direction, respectively.

Figure 4.13 shows the rotation angle ϕ about the x-axis (upper panel) and the torque about the x-axis (lower panel) as a function of time. Apparently the two curves are out of phase to each other but of the same quasi-periodicity. We have seen before in Fig. 4.4 that the rotation accompanies the horizontal movement. Combining these facts, it appears that the swinging pressure generates quasi-periodical torque and horizontal force, which make the ice plate rotate and translate quasi-periodically.

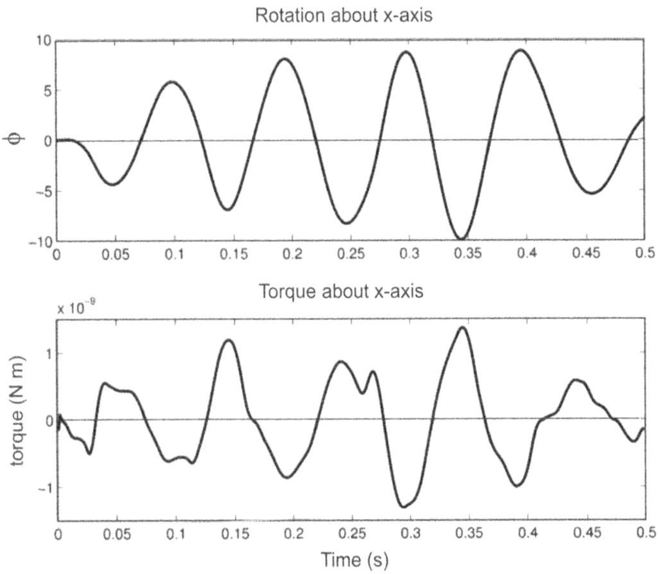

Fig. 4.13 The rotation angle ϕ about the x-axis (upper panel) and the torque about the x-axis (lower panel) as a function of time (adapted from [7] with changes)

This is essentially the same as the analysis of the forces on falling disk by Stringham et al. [46].

By looking at Fig. 4.12, one may get an impression that the main minimum pressure center in the left panel is located to the left side of the crystal. This is not true as the unsteady pressure structure is highly 3-dimensional. A 2-D chart representing a 2-D cross-sectional view may seriously misrepresent the real complex configurations. Figure 4.14 shows the 3-D isosurface low pressure structure corresponding to the two same time frames as in Fig. 4.12. It is clear that the low pressure core tubes (darker blue) spiral left and right as the vortex shedding goes on. Unlike the flow past fixed orientation ice plate case in Wang and Ji [54] where the shedding is purely periodic, the present case allows the ice plate to react to the shedding that modifies the shedding period itself. Consequently, the vortex shedding and the position of the low pressure regions become only quasi-periodic. This and the fact that the oscillation period of the plate is not the same as the vortex shedding period make it difficult to predict the precise location of the vortices at a future time.

Figure 4.15 shows the pattern of vorticity magnitude distribution on the cross-section through the center of falling plate, again corresponding to the same two time steps as in Fig. 4.14. As usual, the vorticity maximum occurs around the edge of the plate and the vorticities are transported downstream by the flow, forming a cup-shaped void that envelopes the wake. Inside the wake, the vorticities are generally low but the pattern is random, and the lowest vorticities occur in the downstream region close, but not immediate, to the plate. Further downstream, the vorticities are not random but form vortex tubes that intertwine together. Figure 4.16 shows

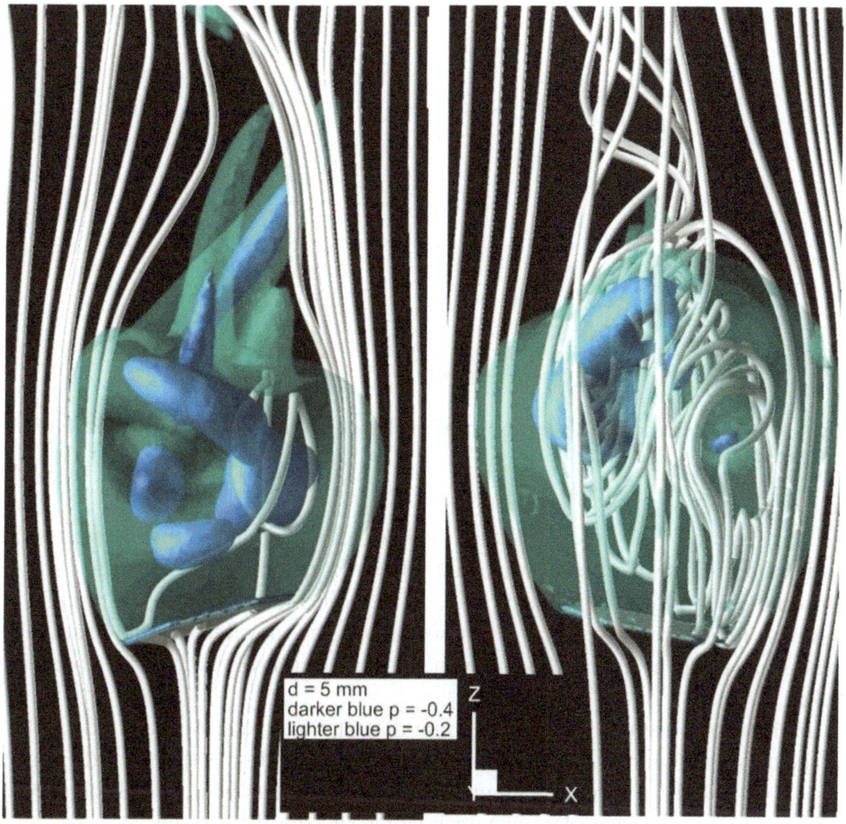

Fig. 4.14 Detailed view of the low pressure isosurfaces in the wake of a falling hexagonal ice plate of 5 mm diameter at two different time steps as in Fig. 4.11

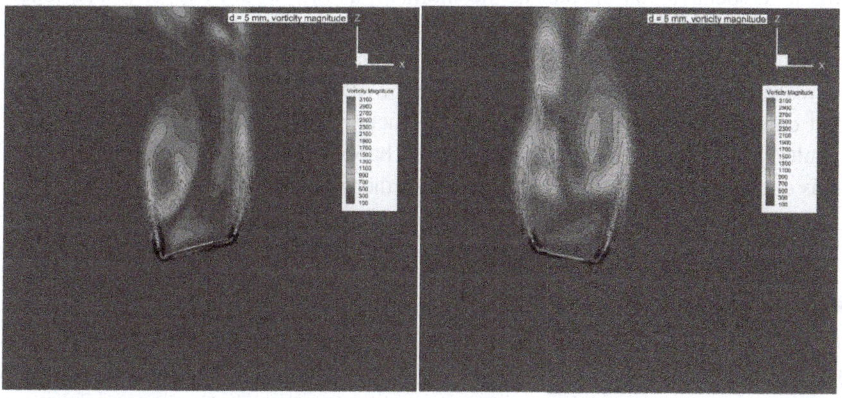

Fig. 4.15 Contours of vorticity magnitude around a falling hexgonal ice plate at time steps as Fig. 4.11 (adapted from [7] with changes)

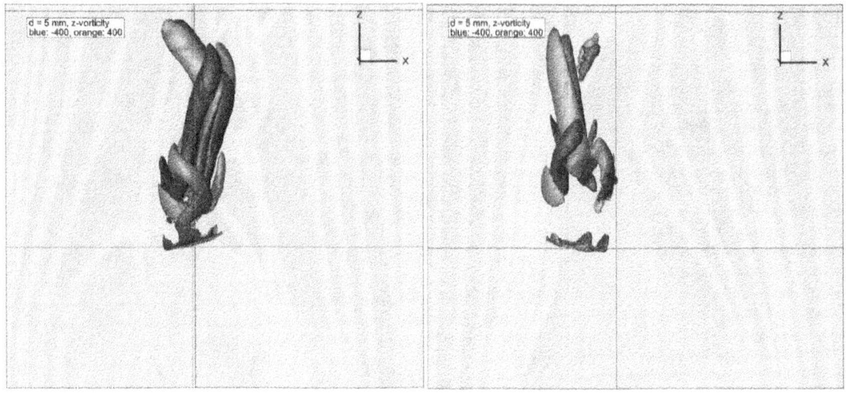

Fig. 4.16 Contour surfaces of z-vorticity around a falling hexgonal ice plate at time steps as Fig. 4.11 (adapted from [7] with changes)

the distribution of the z-vorticity (not total vorticity magnitude) that illustrates the vortex tubes. This wake and downstream flow structure could have important impact on some cloud physical problems such as the collection of tiny cloud droplets (a few μm or less) and scavenging of aerosol particles by snowflakes, as it redistributes these small particles spatially that influences the collection of particles by subsequent flakes.

In the above, we used the 1 and 5 mm diameter ice plates to illustrate the flow characteristics. Figure 4.17 shows a comparison of the flow patterns of all 7 cases ($d = 1, 2, 3, 4, 5, 7, 10$ mm) computed here. It is seen that, while the absolute eddy size increases as the ice plate becomes larger and the flow in the wake becomes more turbulent, the relative eddy length, i.e., the ratio of eddy length-to-plate diameter, seems to be fairly constant for plates with $d \geq 4$ mm. The exact reason of this latter behavior is unclear at the moment, as one would expect to see longer wakes (recirculation bubbles) for larger plates. 2-D steady state simulations by some previous studies (e.g., [45]) did show the very long wakes when the Reynolds number increases. It may be due to the freedom of the plate to oscillate and move sideways that compensates for the necessity of producing longer wakes and, consequently, results in a upper limit of the recirculation bubble size for freely moving bodies? This question deserves to be investigated further in the future.

4.1.5 *Drag Coefficients*

Figure 4.18 shows the time-averaged drag coefficient as a function of the Reynolds number. The drag coefficients from other studies including the experimental data provided by Roos and Willmarth [43] and two numerical results by Wang and Ji [54] and Hashino et al. [14] are also plotted here for comparison. Note that the

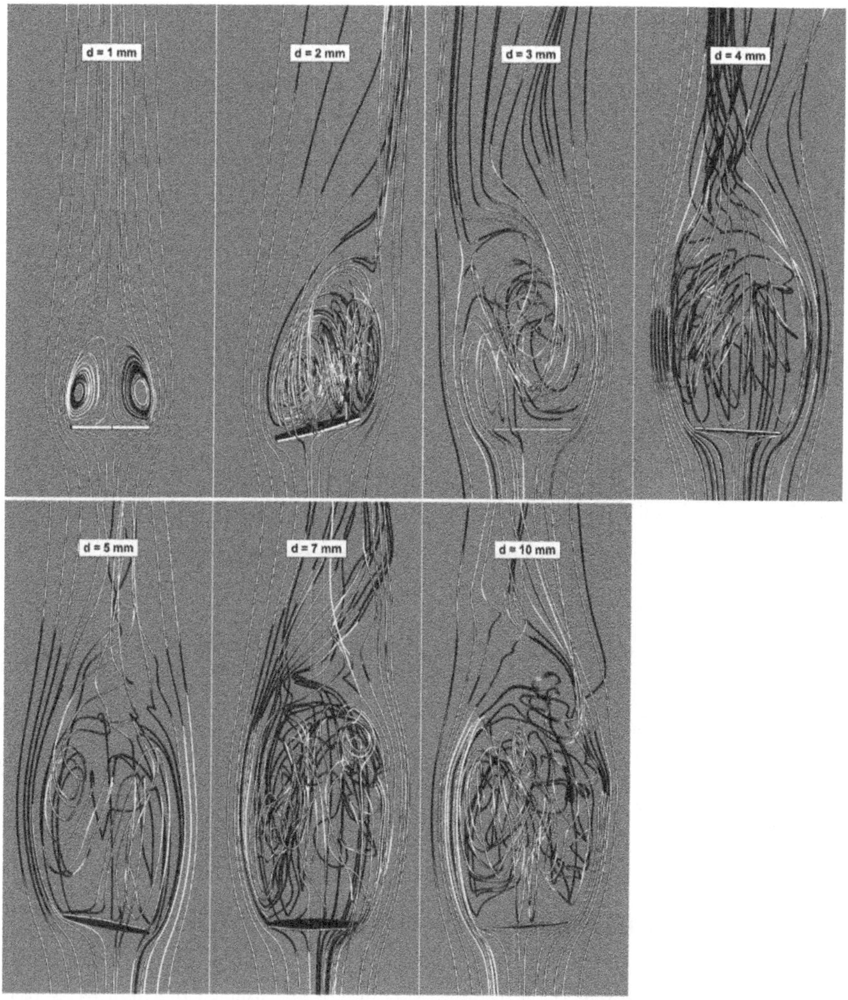

Fig. 4.17 Comparison of the wake patterns of all 7 falling hexagonal ice plates (after [7])

experimental data of Roos and Willmarth [43] is for thin circular disks, not hexagonal plates.

We see that the drag coefficient of a hexagonal ice plate is greater than that of a circular disk at the same Reynolds number. This agrees with the results by List and Schemenauer [28] who compared the drag coefficients of hexagon plates and circular disks based on their laboratory experiments.

The drag coefficient for the hexagonal ice plates in the present study can be fitted by the empirical formula that takes the similar form as that of Wang and Ji [54]:

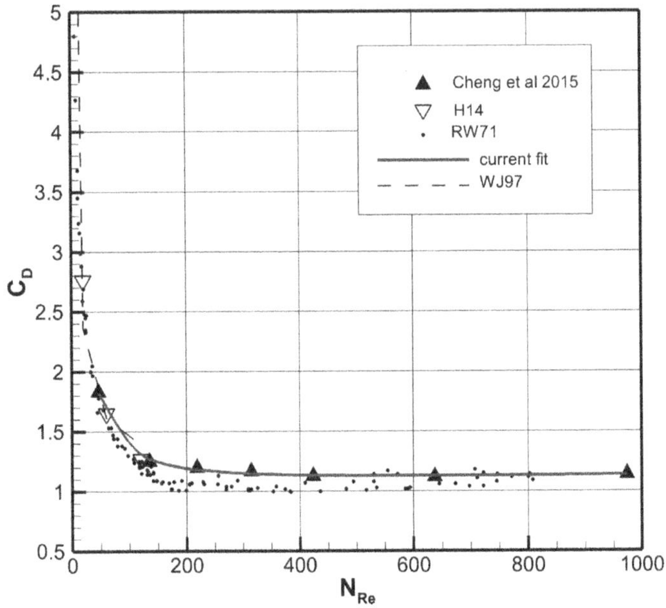

Fig. 4.18 Drag coefficient of falling hexagonal ice plates as a function of Reynolds number (adapted from [7])

$$C_D = \frac{43.48}{N_{\text{Re}}}(1 + 0.01541 N_{\text{Re}}^{1.071}) \tag{4.6}$$

4.2 Flow Fields Around Freely Hexagonal Ice Columns

As we mentioned in Chap. 1, columnar crystals are also one of the major habits in atmospheric clouds. Unlike plates whose largest dimension is along the a-axis, an ice column's largest dimension is along the c-axis of the ice crystal. The drastic difference in habit and shape will naturally lead to different fall behavior. In this section, we will describe the numerical simulation of the free fall behavior of columnar ice crystals. The physics and mathematics are similar to the one we described in Chap. 3 and the previous section for ice plates.

The earliest simulation of the fall behavior of columnar ice crystals was done by Schlamp et al. [44] who approximated an ice column as an infinitely long circular cylinder. Consequently, the simulated flow field can only be 2-dimensional because it is impossible to simulate the effect of the two sharp ends of the column. Also, only steady state flow was examined. Only the two standing eddies of the flow field are somewhat similar to that near the central cross-section of the real flow past an ice column which is obviously of finite length. Ji and Wang [22] performed the first numerical simulation of unsteady flow past a finite-length circular cylinder which is much closer to the real shape of a true ice column. They successfully simulated not only the standing eddies when the larger Reynolds numbers are small but also the eddy shedding phenomenon when the Reynolds number becomes large enough. In addition, they have successfully simulated the pyramidal wake in the downstream of the column as was observed by Jayaweera and Mason [19] in oil tank experiments of the flow around falling short cylinders. While that study represents a great improvement over Schlamp et al. [44] study, it suffers from the assumptions that the orientation of the finite cylinder remains horizontal and its horizontal position remains fixed, i.e., no sideways movement is allowed. Hashino et al. [14] (hereafter H14) investigated the flow fields and hydrodynamic torque acting on inclined planar and columnar crystals but their orientation and horizontal position are still assumed to be fixed. Thus up to that point, the ice column is not really falling freely but is subject to some artificial restrictions. It is not until Hashino et al. [13] that all these restrictions are removed. Hashino et al. [13]'s results are to be summarized below. Their results also contain ice column sizes larger than normally observed in atmospheric clouds but they are included here for a more complete description of hexagonal column fall motions.

4.2.1 Dimensions of Hexagonal Ice Columns

Hashino et al. [13] consider 19 cases of hexagonal columns by varying Reynolds number and aspect ratio (Table 4.3).

The Reynolds number of the ice column is defined as

$$N_{\mathrm{Re}} = \frac{du_\infty}{\nu} \tag{4.7}$$

where u_∞ is its terminal velocity. It is emphasized here that the diameter d instead of the length L is used in the definition of the Reynolds number of the ice column

Table 4.3 Dimensions of the hexagonal columns, resulting terminal velocity and Reynolds number for cases discussed here

Case-ID	$q = L/d$	d (μm)	L (μm)	u_∞ (m/s)	N_{Re^*}
5	3.335	189.9	633.3	0.832	10
10	5	289.9	1449.4	1.311	23.8
20	8.33	472.9	3939.1	2.046	58.4
40	6.29	646.6	4067.2	3.388	139.5

The diameter d and length L are in μm. u_∞ and N_{Re^*} denote the terminal velocity and Reynolds number obtained by the free-fall simulations, respectively (adapted from [13] with changes)

so as to keep in line with the definition used in the past (e.g., [44]) and the results obtained here can be compared directly with previous results.

The moment of inertia around y-axis I_y of a hexagonal ice column is given as

$$I_y = Mr^2\left(\frac{5}{24} + \frac{q^2}{3}\right) \quad (4.8)$$

where $M = 3\sqrt{3}\rho_i r^3 q$ is the mass of the columnar crystal. The dimensions of a hexagonal column and terminal velocity for a given N_{Re} and q are determined from a look up table as was done in H14. The table was constructed by using a general formulation of terminal velocity proposed by Böhm [5]. The N_{Re} specified in this procedure is denoted as $N_{\mathrm{Re}B}$ which varies between 5 and 40, while q ranges from 2.22 to 25.63 as shown in Table 4.3.

An initial perturbation is implemented by allowing the column to begin the fall from an inclined state. As denoted with θ_y in Fig. 3.4, the major (or c) axis of a hexagonal column is rotated around y'' by 20° in counter-clockwise direction, i.e., $\theta_y = -20°$. As in the case of the ice plate flow field calculations, a steady flow field is calculated for the inclined state of the column with a constant U_3. The free-fall simulation is initialized with the flow field and implemented for a dozen time steps. These two steps are repeated with slight changes in U_3, so that the crystal falls by a small amount over the time steps.

Because of the very expensive computational cost, the duration of the free-fall simulation was typically 1 s, but some are 0.5 s. Most of the cases attain a constant vertical velocity, i.e. terminal velocity u_∞, during the period, but some high $N_{\mathrm{Re}B}$ cases do not (see [13]). All the free-fall simulations eventually led to higher u_∞ than those initially calculated with Böhm's formula and the N_{Re} in the free-fall simulation is larger than the specified $N_{\mathrm{Re}B}$ and ranges from 10 to 194 (Table 4.3). We believe that the discrepancy between Böhm's formula and simulated ones is not a problem in the analysis of flow field and stability as the drag and N_{Re} in our calculations are self-consistent.

4.2.2 *Fall Patterns of Hexagonal Columns*

In this section we will describe simulated falling patterns of the hexagonal columns with $N_{\mathrm{Re}} = 10$, 58, and 140 (Cases 5, 20, and 40 in Table 4.3), each represents one of the three characteristic attitudes exhibited by ice columns studied here.

(a) Stable (Damped oscillation)

Figure 4.19 illustrates the damped oscillation as it shows the trajectory and orientation in global x–z plane for the ice column with $N_{\mathrm{Re}} = 10$ (left panel) and the phase plots of its kinematic variables (angle, velocity components) (right panel). We see that the ice column flutters initially but gradually becomes steady and nearly horizontal after

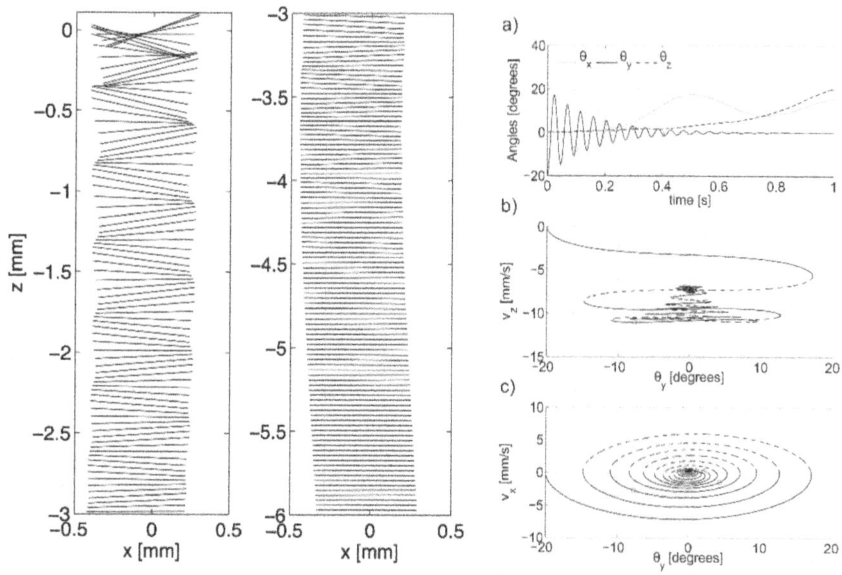

Fig. 4.19 (Left) Damped oscillation mode of a hexagonal ice column falling at $N_{\mathrm{Re}} = 10$. (Right) **a** Tait–Bryan angles versus time; **b** V_z versus θ_y; **c** V_x versus θ_y (adapted from [13] with changes)

z ~ −4 mm. The right-hand panel of Fig. 4.19 shows the kinematic phase plots. We see in Fig. 4.19a that the inclination angle of the *c*-axis relative to the horizontal, θ_y, is strongly damped over the course of falling. The magnitude of vertical velocity of the column relative to the stream flow, v_z, increases initially with the oscillation in Fig. 4.19b but starts to decrease at about 0.1 s along with the amplitude of oscillation. The reduction in the magnitude of oscillation is directly related to the attainment of the terminal velocity since the decrease in θ_y implies the increase in cross sectional area against the flow, therefore the drag increase. When θ_y becomes sufficiently small (after 0.4 s or $z = -3.5$ mm), the column starts a horizontal translation toward positive x direction as shown in the left-hand panel. At this moment, the fall speed reaches the terminal velocity of −0.74 cm s^{-1} relative to the stream flow (Fig. 4.19b). The trajectory in v_x–θ_y space appears as a spiral due to the damping and the ~90° phase difference between them (Fig. 4.19c). The peaks of v_x are achieved when $\theta_y = 0°$ as expected. Note that the horizontal translation is actually larger in y direction than the one in x direction (of global coordinates), and about twice of the length of the crystal at $t = 1$ s. This translation is related to positive rotation along *c*-axis of the column (Fig. 4.19a). At the initial time, a vertex of hexagon of the column points in the negative z direction and the two prism faces adjacent to the vertex lie symmetric across x–z plane. This way horizontal y components of pressure and viscous forces acting on the two prism faces cancel each other. When it rotates around the *c*-axis, the symmetry and the balance in the horizontal forces break down. Thus, the positive (negative) rotation leads to the translation in negative (positive) y direction.

(b) Periodic (Persistent fluttering)

The second mode of fall attitude is persistent fluttering as demonstrated by the hexagonal column of $N_{Re} = 58$ as shown in the left panel of Fig. 4.20. Here the column performs a persistent fluttering in global x–z plane throughout the simulation (Fig. 4.20).

In one cycle of oscillation, the crystal horizontally translates about 1 mm in the x-direction, about ¼ of the length. On the other hand, the y-component of horizontal translation which is ~6 mm at $t = 1$ s is much larger than the x-component, as in the case of $N_{Re} = 10$. This is related to the rotation around the c-axis (Fig. 4.20a). The trajectory in v_z–θ_y space (Fig. 4.20b) assumes a butterfly shape, similar to that reported by Belmonte et al. [3] for a falling paper in dense liquid. The shape basically indicates that the phase of v_z is shifted by more than 90° and the frequency is about half of θ_y. The slowest v_z is achieved at around ±13°, not at 0°. The fastest v_z is achieved at around ±16°, not at the largest inclination angle. The v_z keeps oscillating, but reaching a quasi-steady state after 0.5 s. The fluctuation in v_x (~70 mm s^{-1}) is

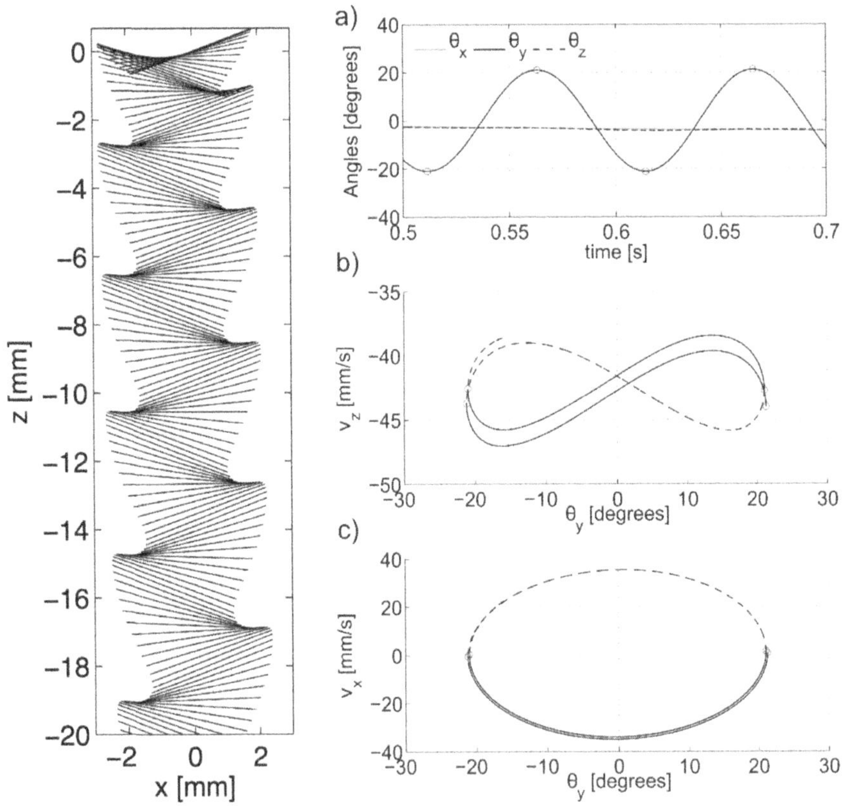

Fig. 4.20 (Left) Periodic mode of a hexagonal ice column falling at $N_{Re} = 58$. (Right) **a** Tait–Bryan angles versus time; **b** V_z versus θ_y; **c** V_x versus θ_y (adapted from [13] with changes)

about ten times larger than one in v_z (~7.5 mm s^{-1}). Since the damping is negligible, the trajectory in v_x–θ_y space is close to an ellipse (Fig. 4.20c).

(c) Unstable (Tumbling)

The third model of ice column fall attitude is the unstable mode as illustrated by the left hand panel of Fig. 4.21 for the hexagonal column of $N_{Re} = 139$. Here it shows that the column falls in an unstable manner where the magnitude of oscillation angle θ_y keeps increasing as it falls (Fig. 4.21a).

The fall velocity is positive for the first 0.2 s due to the large stream velocity. As θ_y increases, the fall velocity also increases (Fig. 4.21b) due to the reduction in cross-section area and hence the drag. After it reaches z ~ −80 mm at about 0.6 s, it starts to tumble in the global *x*–*z* plane. Once it reaches 90°, the crystal quickly rotates around the *c*-axis by about 180° before the magnitude of θ_y starts decreasing (as indicated by θ_x of Fig. 4.21a). This leads to the *apparent* tumbling in global *x*–*z* plane although the crystal keeps oscillating around the local *y* axis. The amplifying oscillation of θ_y is accompanied by increase in v_x, which assumes an outward spiral motion in v_x–θ_y space (Fig. 4.21c). The amplitude of oscillation in v_x reaches to more than 10 cm s^{-1}, which is accompanied by the *x*-translation of about 45 mm at

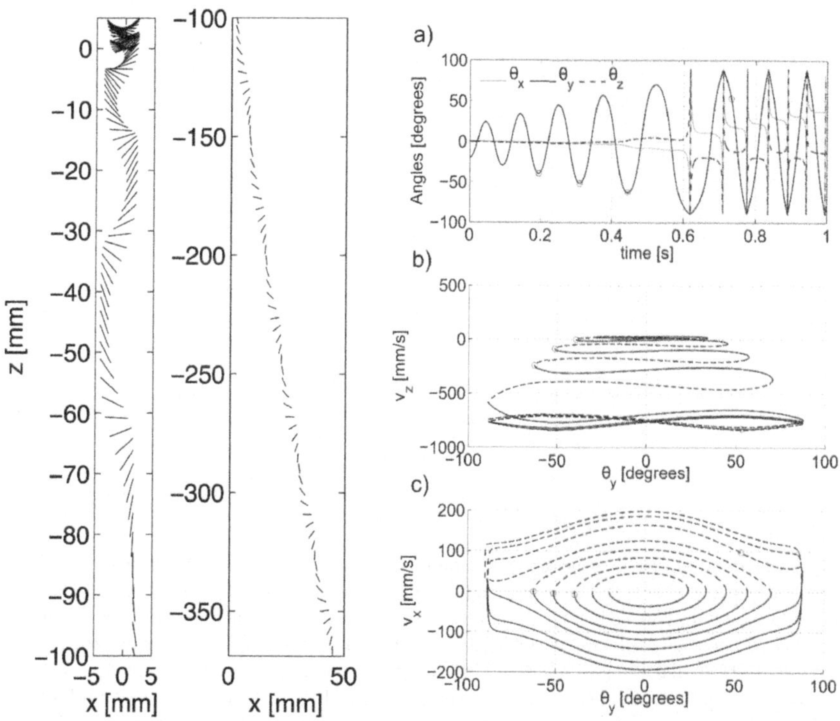

Fig. 4.21 (Left) Unstable mode of a hexagonal ice column falling at $N_{Re} = 139$. (Right) **a** Tait–Bryan angles versus time; **b** V_z versus θ_y; **c** V_x versus θ_y (adapted from [13] with changes)

t = 1 s. This is more than ten times of the column length. After about 0.732 s, v_z of the column oscillates at ~ −75 cm s^{-1} (Fig. 4.21b), meaning that it has reached a steady state with terminal velocity.

4.2.3 Stability Diagram

The above discussions illustrate that ice columns can fall in three modes: stable, periodic and unstable. The fall mode depends on N_{Re}, the aspect ratio (d/L), and torque. Figure 4.22 shows the stability diagram of the falling columns examined here. It is seen that there is a bow-shaped boundary, indicated by the thick dashed curve, that separates the stable and unstable regimes. The unstable mode regime is the in the upper right region (with light grey filled symbols) of the chart representing generally higher N_{Re}. The rest of the region represents the stable regime (dark filed symbols, N_{Re} < ~50 or when d/L × 100 < ~4 for N_{Re} < 200). This is very similar to the finding of [19] (see their Fig. 7) who showed that there is a boundary between a steady motion regime without oscillation (or fluttering) and persistent fluttering

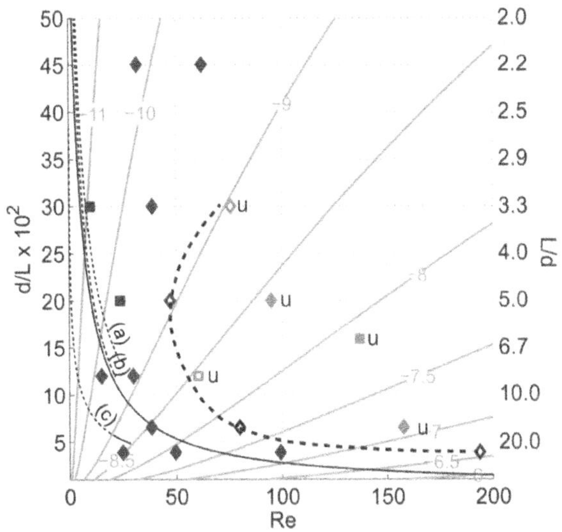

Fig. 4.22 Stability of falling hexagonal columns in terms of Re and inverse of aspect ratio. The right vertical axis indicates the corresponding aspect ratio $q = L/d$. Those marked with an "u" is unstable and the rest are stable. Open symbol represents borderline condition. The dashed curve is a boundary between stable and unstable motion. The gray contours show constant torque factors [$\log_{10}$(N m)] calculated with Eq. (9) of [14]. The black curve shows a power-law fit to columns observed by Auer and Veal [2], extended beyond the limit of Re = 20. Dashed curve **a** is for rimed long columns of Mitchell [32], **b** is for hexagonal columns of Mitchell [32], and **c** is for elementary needles of Mitchell et al. [34]. The Re was calculated with the formula of Heymsfield and Westbrook [17] (adapted from [13] with changes)

regime. At the boundary the particles were observed to oscillate but with heavy damping. Even though JM65 studied steel cylinders falling in liquid paraffin, our results match with theirs in terms of N_{Re} and q^{-1}. A new finding of this research is that the stability boundary appears to curve toward larger N_{Re} with d/L from $N_{Re} = 50$ and $d/L \times 100 = 20$, or in other words the stable area expands to larger N_{Re} as q decreases from 5.

Also shown in Fig. 4.22 are ice column size distributions studied by some previous researchers. The black curve represents a power-law fit to columns observed by Auer and Veal [2] which is extended beyond the sample limit of $N_{Re} = 20$. The orange and green curves represent that the long columns and hexagonal columns by Mitchell [32] respectively and the violet curve represent the elementary needles by Mitchell et al. [34]. Clearly these examples all fall in the stable regime. In all previous studies, the N_{Re} of hexagonal columns (including rimed columns and needles) are all less than 60 and the aspect ratio less than 25 [24, 59], so generally they never grow large enough to enter the fluttering or unstable modes. Therefore, we may conclude that pristine columnar crystals in the atmosphere generally fall in the stable mode possibly reach the fluttering regime, but unlikely to fall in the unstable mode.

4.2.4 Torque and Flow Fields

Given the importance of rotational motion in the fall attitude of ice columns, it is of interest to examine the issue of torque in more detail. The torque acting about local y axis can be decomposed to those acting on faces of the crystal. We shall name the pair of prism faces in positive (negative) z direction as Prism $z+$ (Prism $z-$) at the initial time (Fig. 4.23).

The torque can be divided into two parts: pressure and viscous torques. The pressure torque acting on Prism $z+$ with respect to the local y axis (or y'' axis) can be calculated as a surface integral of the y'' component of the torque:

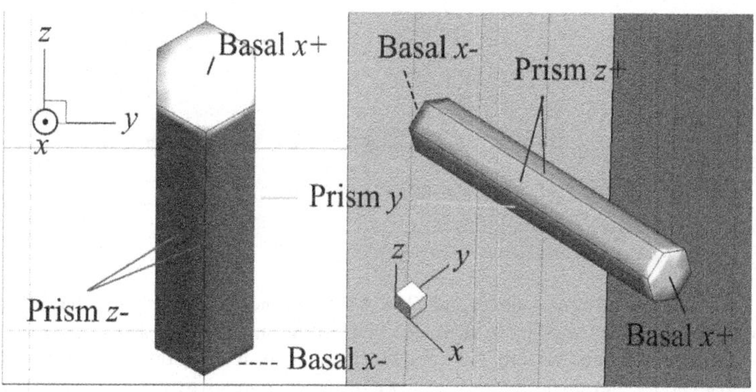

Fig. 4.23 Nomenclature of column geometry for examining the torque (adapted from [13] with changes)

$$T_{p,y} = \oint \left\{ (\vec{x}_p - \vec{x}_{cg}) \times p_s \vec{a} \right\} \cdot \overrightarrow{y''} dA \tag{4.9}$$

where $\vec{x}_p$ is the position vector on the face, $\vec{x}_{cg}$ the center of gravity of the column, $\vec{a}$ the normal vector of the face, and $\overrightarrow{y''}$ the unit vector of local y axis in global coordinates. For the viscous torque, $p_s\vec{a}$ is replaced with viscous stress acting on the face.

In the lower Reynolds number case of $N_{\mathrm{Re}} = 10$, the contribution pressure torque to the total torque is about 56% which indicates that the viscous torque also has substantial contribution. For the case of periodic fall mode ($N_{\mathrm{Re}} = 58$), the pressure contribution to the total torque becomes larger (71%). But when the fall motion becomes unstable (Re = 139), most of the torque (about 81%) is created by the pressure. A detailed analysis of torque acting on these columns is given by Hashino et al. [13].

Figure 4.24a, d show the flow around the column at $N_{\mathrm{Re}} = 10$. The streamtraces *show that the flow is laminar and there is no standing eddy attached to the column.* This is similar to the steady-state case of $N_{\mathrm{Re}} = 2$ discussed in H14. The shades that show the non-dimensional horizontal velocity. The fact that the pressure and viscous torques on Prism z− are not zero at $\theta_y = 0$ suggests that the K and NV_y are

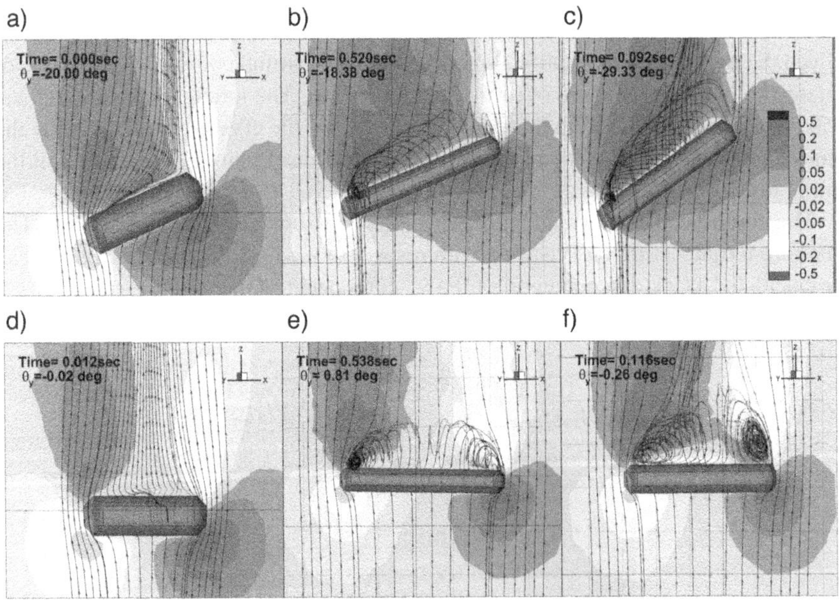

Fig. 4.24 Examples of stream traces and flow fields around the hexagonal column (top) with and (bottom) without inclination: **a**, **d** for Re = 10; **b**, **e** for Re = 58; and **c**, **f** for Re = 139. Nondimensional horizontal velocity (air velocity along the local x axis divided by u′) is shown by the shaded contours, and stream traces are shown by the gray lines with arrows (adapted from [13] with changes)

not exactly symmetric across y–z plane. However, they are quasi-symmetric. This suggests that the steady state solutions obtained in H14 can be used as approximations when rigorous but computing expensive solutions are not available.

The periodic fall mode ($N_{\mathrm{Re}} = 58$) accompanies unsteady eddy formation behind the column (Fig. 4.24b, e). When the $\theta_y \sim 0°$, the stream traces from the both upper corners appear to meet almost at the center of the crystal (Fig. 4.24e). When θ_y increases from -18.4 to $0°$, the crystal translates to negative x direction. The flow pattern is again quasi-symmetric.

The flow fields for the unstable fall mode ($N_{\mathrm{Re}} = 139$) appear to be qualitatively similar to the periodic case ($N_{\mathrm{Re}} = 58$) at the first glance (Fig. 4.24c, f). However, since Re is higher, the length of the eddy in z direction, the magnitude of horizontal velocity are larger than that of $N_{\mathrm{Re}} = 58$. The distinct difference here from the periodic and stable cases is that the flow separation occurs in the middle of the right basal face. This leads to the zero viscous torque acting on Basal $x+$. When the column becomes horizontal, the asymmetry in the eddy is more pronounced than that in $N_{\mathrm{Re}} = 58$. This asymmetry in the eddy circulation behind produces the driving torque toward unstable motion.

4.2.5 *Drag Coefficients*

The drag coefficients C_D for all the cases were calculated using the fall velocity and drag when the inclination angle θ_y is zero near the end of each simulation (consistent with definition of N_{Re}). Figure 4.25 shows a comparison of the C_D obtained from the present free-fall simulation with those obtained by previous investigations. First, it is clear that for steady-state simulation C_D for hexagonal columns (open triangles) is larger than one for circular cylinders (open circles) for the same N_{Re} and q. Secondly, the steady-state C_D for hexagonal columns agrees well with a semi-empirical parameterization of Heymsfield and Westbrook [17] for columns and rimed long columns (thick solid and dot-dash gray curves), which validates the relationship between Re and C_D obtained from Böhm's [5] general formula for steady-state simulations. Finally, the simulated C_D (filled squares and diamonds) also matches well with the C_D of Heymsfield and Westbrook. This supports the results of numerical simulation obtained under free-fall settings.

Surprisingly, the parameterization given by Wang and Ji [54] for hexagonal plates (thin solid curve) matches well with the samples obtained for hexagonal columns, irrespective of the stability. Thus, the parameterization can be used for hexagonal columns, which is

$$C_D = \left(\frac{64}{\pi N_{\mathrm{Re}}}\right)(1 + 0.078 N_{\mathrm{Re}}^{0.945}) \tag{4.10}$$

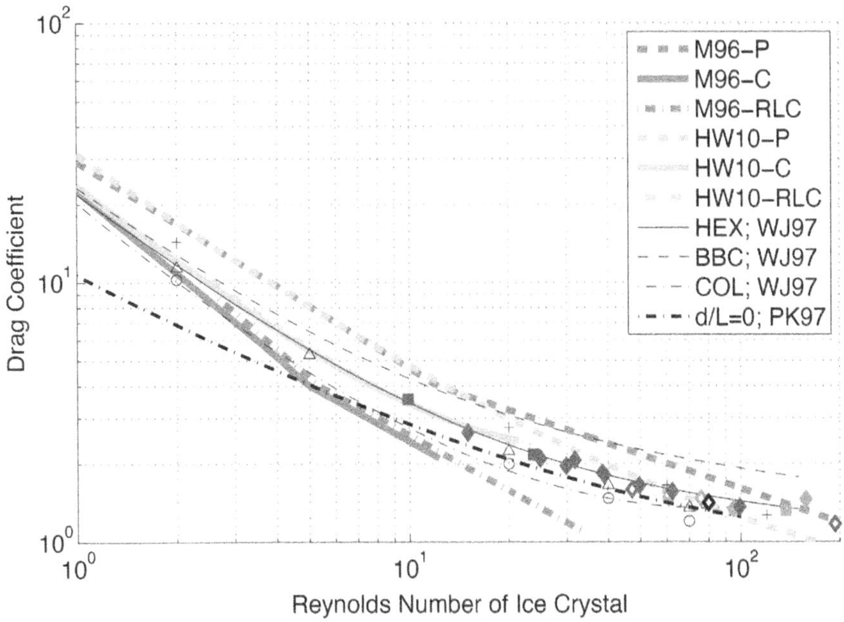

Fig. 4.25 Comparison of drag coefficient of hexagonal columns against previous parameterizations. The filled squares and diamonds are from the free-fall simulations. The drag coefficients from steady state solutions from Hashino et al. [14] are shown: circles for cylinders, crosses for hexagonal plates, and open squares for hexagonal columns. Thin lines—parameterizations of Wang and Ji [54]: solid curve for hexagonal plates, dash curve for broad-branched crystals, and dot-dash curve for columns. Thick dash-dot—parameterization for infinite-long circular cylinder by Pruppacher and Klett [41]. Thick dark grey—Mitchell [32], Thick light gray—Heymsfield and Westbrook [17]: dash for hexagonal plates, solid for hexagonal columns, dot-dash for rimed long columns (adapted from [13] with changes)

As discussed in H14, C_D by Wang and Ji tends to be smaller than the experimental data. This point is also clear by comparing the parameterizations by Mitchell [32] and Heymsfield and Westbrook [17] for hexagonal plates (thick dash curves).

4.2.6 *Impact of Fluttering and Rotation of Ice Columns*

As we have seen above, persistent fluttering of ice columns may occur in the atmosphere. This fluttering fall attitude would affect the collection of other particles, for example, small supercooled droplets, by the column, similar to what we have described previously (see Fig. 4.4). For columns, we use the case $N_{\mathrm{Re}} = 58$ to illustrate this effect. Since the inclination angle θ_{yF} relative to the fall velocity vector oscillates with larger amplitude than θ_y, the fluttering leads to the reduction of the cross sectional area projected perpendicular to the particle motion vector. This cross-sectional area sweeps out a volume per unit time which is called the geometrical kernel K^*, and important quantity in defining the collision efficiency of hydrometeors. The collision efficiency E is defined as $E = K/K^*$ where K is the collision kernel which is the volume within which a particle will be collected by this hydrometeor (see [50, 57]). Thus by changing K^* the efficiency E will also be changed. When a cloud droplet with zero fall velocity is situated below the fluttering column, the reduction in sweeping volume relative to the conventional sweeping volume (i.e., the collection kernel defined as the constant terminal velocity times cross section area without inclination. See Wang [50, 57] is only 1.56%. As the fall velocity of cloud droplets increases, the contribution of horizontal component of fall velocity increases relative to the fall velocity difference, but even if the droplet is moving at 99% of the fall velocity of column, the reduction is only 7.1%. Therefore, the impact of persistent fluttering on sweeping volume and hence the collision growth is likely to be small.

4.3 Stellar and Broad Branch Crystals

In the previous sections, we discussed the fall behaviors of hexagonal ice plates and ice columns, both are the most fundamental habits of ice crystals. In this section, we shall present the theoretical numerical solutions of the flow fields around ice crystals of more complex shapes: four types of planar ice crystals—crystals with sector-like branches (P1b), crystals with broad branches (P1c), stellar crystals (P1d), and ordinary dendrites (P1e) for crystal size ranges of 0.2–5 mm in maximum dimension.

As we know that the dimensions (diameter, thickness) of ice crystals of a certain habit are not arbitrary but follow certain empirical relations. For the present study, the ice crystal diameter and thickness are determined using empirical power law relationships given by Auer and Veal [2]. The general atmospheric and microphysical properties are the same as those reported in the precious sections. The crystal density ρ_s is assumed to be 916.68 kg m^{-3}. The plates are assumed to fall in an atmospheric environment of air density $\rho_a = 1.19$ kg m^{-3} and kinematic viscosity $\nu = 1.4019 \times 10^{-5}$ m^2 s^{-1} (approximating environmental conditions 900 hPa and -10 °C). The complete results are given in Nettesheim and Wang [35].

4.3.1 Dimensions of Planar Ice Crystals

In this section, we report the results of falling stellar and broad branch crystals (BBC) with diameters (maximum dimension) 0.2, 0.3, 0.4, 0.5, 1, 2, 3, 4, and 5 mm. The ice plates are not assumed stationary initially, rather they fall at a reference velocity, close to the terminal velocity. Velocity power law relationships from Mitchell and

Heymsfield [33] are used for determining a first guess terminal velocity. Most simulations were repeated several times using refined reference velocities to improve convergence and reduce boundary condition influences.

We have performed numerical simulations of 36 cases of ice crystals but the results are comparable across different crystal sizes, and thus we only report a select number of cases in the following.

4.3.2 *Flow Characteristics*

Figure 4.26 shows the pressure distribution and streamlines around two planar ice crystals with Reynolds number near 150 (range 115–175 corresponding to the 2 mm sector plate *P1b* and 3 mm broad-branched crystal *P1c*) on the y–z plane for a randomly selected time step consistent across each panel.

The diameter d used in Reynolds number $N_{\mathrm{Re}} = (du_{\infty})/\nu$ is that of the maximum dimension of the crystal. The flows are steady in each case at this lower Reynolds number range, with a general left-right symmetry about the central x axis, despite slight rotation about the z axis in many cases. The region of relatively higher pressure in the upstream region is expected. Standing eddies are present in the wake of the sector plate (P1b) and broad-branched crystal (P1c); the ice plates are falling vertically, while the characteristics of the eddies are steady state, i.e., they remain constant with time. The ice crystals mainly affect the flow within the vertical column in which they fall, only slightly disturbing the area outside this column up to a distance of a few diameters. The results are consistent across the current study for crystals with Re near and less than 100, and is in general agreement with previous studies (e.g., [7, 58]).

While negative deviations in pressure generally occur in the downstream wake, pressure minima also occur at the upper edge of the plate (Figs. 4.26 and 4.27).

This behavior is consistent with other numerical studies which treat planar ice crystals as hexagonal plates with finite thickness (two sharp edges), such as Cheng et al. [7], Hashino et al. [14], Ji and Wang [22], Wang and Ji [54]. This configuration of the pressure field at the upper edge of an ice plate likely impacts the collision efficiency, altering the riming rate and the location at which a supercooled drop will strike a plate. This pressure configuration is consistent across all plates in this study.

Figure 4.28 shows the pressure distribution over the surfaces of a 2 mm broad-branch crystal (P1c). It demonstrates that the lowest pressure regions are distributed around the edges of the branches in the downstream side while the high pressure is concentrated at the center in the upstream side, and both high- and low-pressure distributions maintain a hexa-symmetry as they should be because the flow is steady.

This configuration works to explain the process of growth along branches, reproducible in modeling studies (e.g., [12, 37]). The pressure distribution on the surface of the high pressure side of the crystal is more chaotic than the other side. An extremely small sliver of low-to-negative pressure is observed along the entire perimeter of the branches. Finally, the side-view echoes the observations from the basal surfaces;

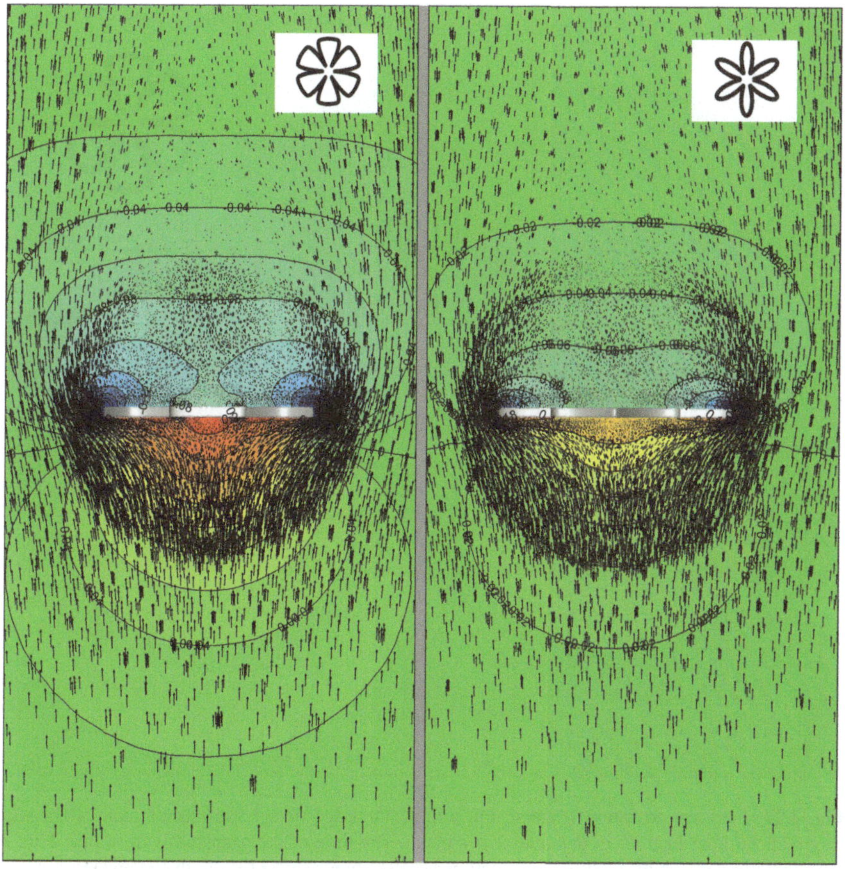

Fig. 4.26 Pressure deviation (Pa) and streamlines (2D projection onto the central y–z plane) around planar crystals (left: 2 mm P1b; right: 3 mm P1c) with Reynolds number near 150. Pressure is shown by color shades (red: positive; blue: negative) and contoured (solid: positive; dashed: negative) over a range of −0.35 to 0.35. All snapshots correspond to the same randomly selected time step

the pressure distribution is such that the minima occur at the peaks, with pressure increasing towards the center of the crystal.

The flow becomes unsteady with increasing Reynolds number as expected as shown in Fig. 4.29 where d = 5 mm and Re = 350. The stellar crystal (P1d) and ordinary dendritic crystals (P1e) do not exhibit unsteady fall behavior over any range of Reynolds number simulated in this study; the sector plate (P1b) and broad-branched crystal (P1c) begin to demonstrate unsteady behavior for diameters of 3 mm and above for the sector plate, and 4 mm and above for the broad-branched crystal, corresponding the Reynolds numbers 198 and 258, respectively. Snapshots of a 5 mm diameter sector plate and broad-branched crystal are presented in Fig. 4.29. The left-right asymmetry of the flow patterns is obvious. The upstream high pressure region,

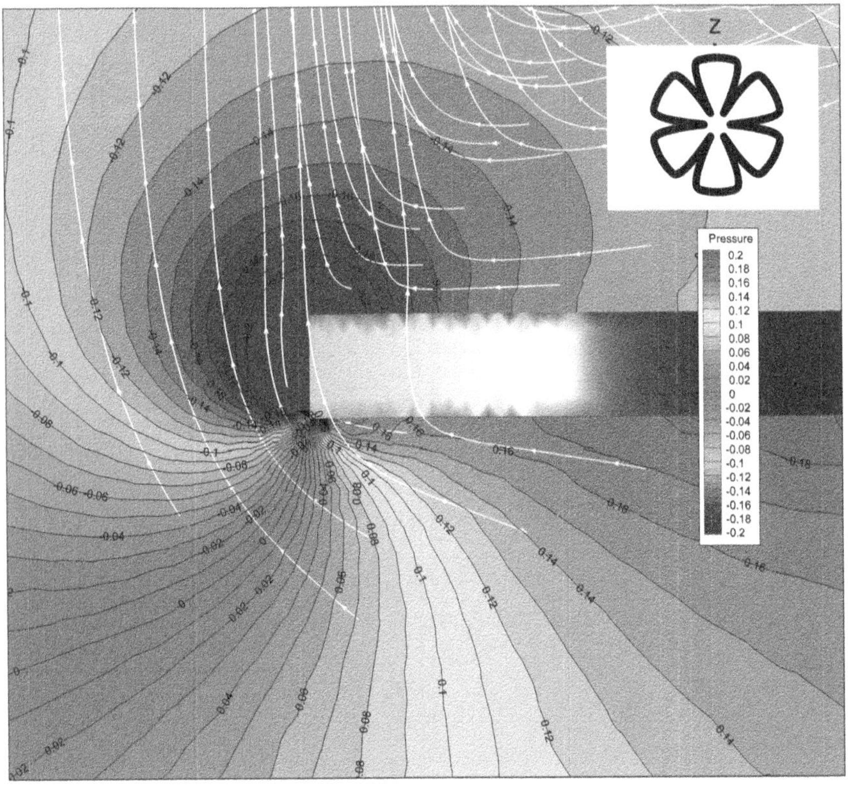

Fig. 4.27 Pressure deviation (shaded, Pa) around the edge of a 1 mm sector plate (P1b) with Reynolds number near 40 in the central y–z plane. The time of the snapshot is as in Fig. 4.26. A few streamtraces (white) on this cross-section are also shown (after [35] with changes)

along with pressure minima along the edge of the crystal surface, are two similarities between the steady cases.

The eddies in the downstream wake are no longer symmetric, with eddy shedding occurring in the downstream. Unlike the steady cases in Fig. 4.26, the pressure minimum in the wake is not always in contact with the surface of the plate, but is found slightly above the surface, or in contact with only one region. The pressure minimum changes in both magnitude and location on a near-periodic basis, resulting in eddy shedding. Consistent with Cheng et al. [7], the location of the upstream pressure maximum is not stationary either, but tends to vary with the orientation of the crystal. As the crystal tilts in a direction, the asymmetric maximum pressure distribution on the underside of the plate will generate a torque, causing the crystal to begin to take on the opposite inclination and move in the opposite direction.

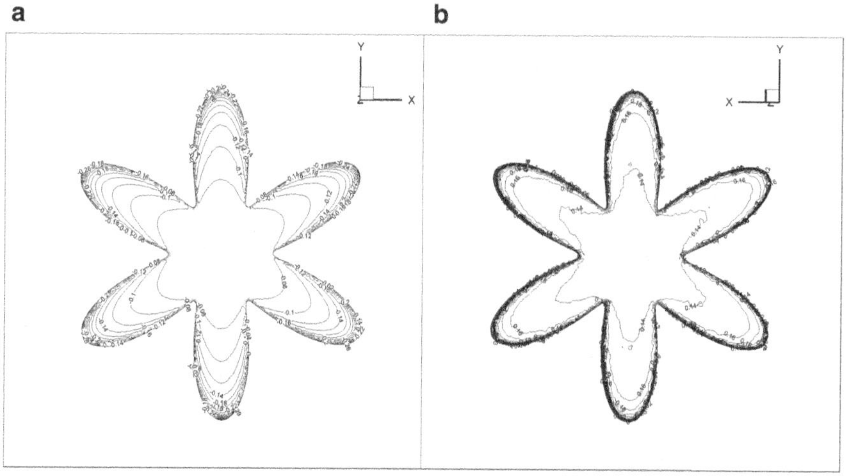

Fig. 4.28 Pressure deviation (shaded, Pa) on the top (**a**), bottom (**b**), and side (**c**) surfaces of a 2 mm broad-branched crystal (P1c) with Reynolds number near 95. The time of the snapshot is as in Fig. 4.26 (adapted from [35])

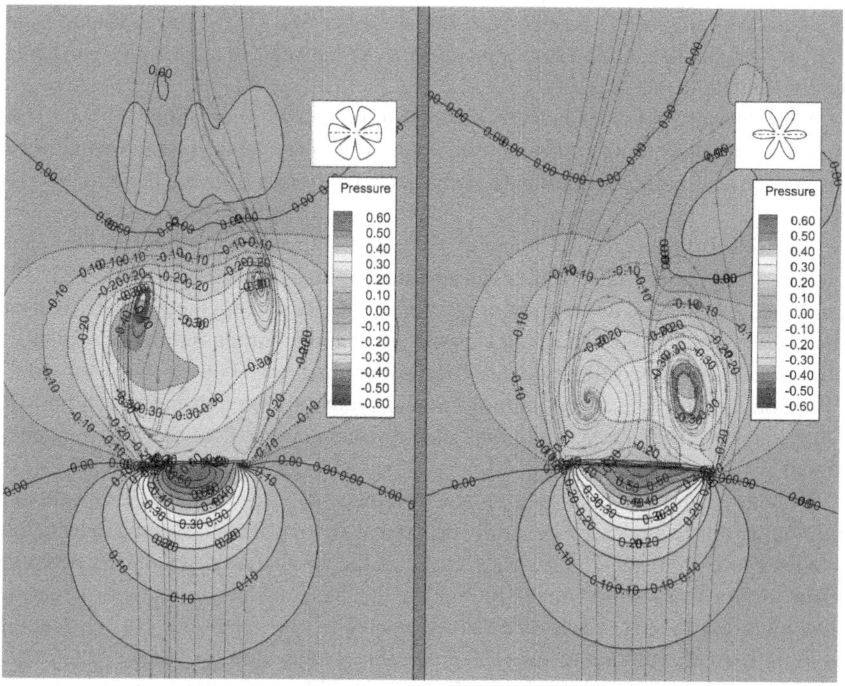

Fig. 4.29 Pressure deviation (Pa) and streamlines (2D projection onto the central y–z plane) around planar crystals (left: 5 mm broad-branched crystal *P1c*; right: 5 mm sector plate *P1b*) with Reynolds number near 350. Pressure is shown by color shades (red: positive; blue: negative) and contoured (solid: positive; dashed: negative) over a range of −0.60 to 0.60 (after [35] with changes)

4.3.3 Fall Attitudes

An analysis of the dimensionless displacement, vibration frequencies, and characteristic angles is performed, and the results of each crystal type are presented in Table 4.5. Results are omitted when the vibration frequency or characteristic angles are not obvious or would require subjective interpretation. The vibration frequency (frequency of the periodic behavior) about the x and y axis are generally the same, though it is noted when the frequencies differ slightly, and both frequencies are listed for 0.5 mm ordinary dendritic crystal (P1e), which differed significantly.

The dimensionless horizontal displacement l^* describes the characteristic length of the horizontal motions relative to the largest dimension of the ice crystal, defined as:

$$l^* = \frac{l}{d}, \tag{4.11}$$

where l is the length of swing, spiral and irregular motions of the crystal. The smallest dimensionless horizontal displacement of the crystals is 0.009 while the largest is 3.525. The crystal type with the largest range in l^* is the sector plate (P1b); the smallest range in l^* is seen with the stellar crystal (P1d). As in Cheng et al. [7], no systematic relationship appears to exist between diameter and horizontal dimensionless displacement.

As mentioned previously, the horizontal motions and orientation of ice crystals may have impacts on the collision efficiency of two crystals. Crystal orientation is also important when considering radar backscatter signals from ice plates (e.g., [18, 31]).

The Tait–Bryan angles, φ, θ, and ψ, are used to quantitatively describe the orientation of the ice crystals with respect to their initial orientation (Table 4.4). The angles φ, θ, and ψ are defined and correspond to rotation about the x, y, and z axes of the crystal, respectively. The characteristic angle and vibration frequency results are confirmed using supporting evidence from performing a fast Fourier transform (FFT) of the orientation data. The vibration frequency about the z axis have been omitted when the high-frequency signal was not obvious compared to the low-frequency signal.

Rotation about the z axis is observed with the small, steady state crystals, though the characteristic angles are generally ordered at 10^{-1} or $10^{-2\circ}$. Nothing in principle should cause spinning in the steady cases of the idealized crystals. This behavior likely arises from slight imperfections in the meshing processes, which causes inevitable asymmetry. Lack of exact symmetry of the mesh and crystal, leading to an uneven distribution of mass, is a possible explanation for some crystal cases demonstrating oscillation about a preferential basal face-parallel axis (a-axis), even though the moments of inertia about any a-axis are the same.

The x–z plane trajectories for crystals with Reynolds number near 350 are shown in Fig. 4.30.

Table 4.4 Dimensionless displacement, vibration frequencies and characteristic angles for the planar ice crystals investigated in this study

Type	Diameter (mm)	$l^* = l/d$	Vibration frequency along x and y axis (Hz)	Vibration frequency along z axis (Hz)	Characteristic φ (°)	Characteristic θ (°)	Characteristic ψ (°)
P1b	0.2	3.525	3.66	1.22	0.12	0.12	0.14
	0.3	0.510	25.64	12.21	0.30	0.19	0.05
	0.4	0.052	30.52	7.32	0.11	0.04	0.03
	0.5	1.000	29.30	8.55	0.09	0.17	0.01
	1	0.235	25.64	18.81	0.07	0.17	0.27
	2	0.069	17.09	–	1.23[a]	0.24[a]	–
	3	0.826	13.43	4.88	1.12	1.49	0.01
	4	0.115	10.99	7.32	2.68	1.17	0.01
	5	0.263	8.55[b]	23.19	1.77	3.15	0.27
P1c	0.2	0.043	13.43	13.43	0.01	0.01	0.03
	0.3	0.015	4.88	3.66	0.02	0.02	0.12
	0.4	0.761	1.22	4.88	0.16	0.09	1.16
	0.5	2.466	1.22	3.66	0.09	0.11	0.07
	1	0.015	24.41	–	0.01	0.04	–
	2	0.009	19.53	–	0.01	0.01	–
	3	0.012	14.65	–	0.24[a]	0.28[a]	–
	4	0.054	12.21	1.22	0.88	0.29	0.08
	5	0.052	10.99	1.22	4.30	1.97	3.45
P1d	0.2	0.059	–	14.65	–	–	0.00
	0.3	0.274	3.66	1.22	0.39	0.82	0.03
	0.4	0.199	3.66	25.64	0.58	0.44	0.01
	0.5	0.217	3.66	8.55	0.23	0.42	0.01
	1	0.141	10.99	3.66	0.07	0.13	0.14
	2	0.162	4.88	–	0.14	0.21	–
	3	0.310	3.66[b]	–	0.01	0.59	–
	4	0.121	3.66	3.66	0.96	0.30	0.02
	5	0.227	3.66	–	0.57	0.53	–
P1e	0.2	0.060	4.88[b]	10.99	0.09	0.04	0.02
	0.3	0.554	1.22	–	0.09	0.06	–
	0.4	0.796	7.32	–	0.04	0.06	–
	0.5	0.415	1.22/4.88	8.55	0.08	0.15	0.01
	1	0.293	3.66[b]	–	0.00	0.02	–

(continued)

Table 4.4 (continued)

Type	Diameter (mm)	$l^* = l/d$	Vibration frequency along x and y axis (Hz)	Vibration frequency along z axis (Hz)	Characteristic φ (°)	Characteristic θ (°)	Characteristic ψ (°)
	2	0.120	10.99	3.66	5.32[c]	1.06[c]	8.38
	3	0.144	7.32[b]	3.66	1.92[c]	3.77[c]	3.16
	4	0.047	4.88	–	0.04	0.06	–
	5	0.044	4.88[d]	–	0.03	0.01	–

[a]Amplitude shows an increasing trend with time, may be underestimated
[b]The vibration frequency about the x axis is slightly larger than the y axis
[c]Amplitude shows a decreasing trend with time, may be overestimated
[d]The vibration frequency about the y axis is slightly larger than the x axis
The dimensionless displacement describes the characteristic length of the horizontal motions relative to the largest dimension of the ice crystal. The vibration frequency is defined as the frequency of the periodic behavior. The characteristic Tait–Bryan angles, φ, θ, and ψ, correspond to rotation about the x, y, and z axes of the crystal, respectively, and describe the orientation of the ice crystals with respect to their initial orientation (adapted from [35] with changes)

The starting positions, with no initial inclination angle, are shown by the projections marked with thick black bars, and so are the end positions, occurring at 0.5 s. The projections are displayed every 0.005 s. These cases become less stable with time, and appear quite unstable towards the end of the simulation time. The zigzag swing oscillations in the sector plate (P1b) result in horizontal translations of roughly 26% of diameter of the plate; fluttering in the broad-branched crystal (P1c) produces a smaller horizontal displacement of 5% of its diameter. Intuitively, the crystal is displaced rightward (leftward) horizontally in response to tilting to the right (left), in agreement with experimental results of Willmarth et al. [58] and Stringham et al. [46] and consistent with the numerical results of Cheng et al. [7] and Hashino et al. [13].

An observational study by Kajikawa [25] showed a relationship between increasing Reynolds number, above about 40, and unstable falling motion due to vortex shedding. In that study, unstable falling motion first began with oscillations about an *a*-axis of plate-like crystals, then proceeded to display swinging motion, followed by rotation about the z axis, with increasing N_{Re}. Note, tumbling motions that were observed in tank experiments by Willmarth et al. [58] and Stringham et al. [46] were not observed in the Kajikawa [25] observational study. Further, due to natural snow crystal asymmetry, the onset of observed unstable falling motion in the Kajikawa [25] study occurred at considerable smaller *Re* values than in model experiments (e.g., [28]). In the current CFD model experiment, the onset of unstable fall motion did not occur until higher N_{Re}, with the highest N_{Re} for stable motion occurring at Reynolds numbers 115 and 175 for the simulated sector plate (P1b) and broad-branched crystal (P1c), respectively; the stellar crystal (P1d) and ordinary dendritic crystal (P1e) exhibits stable falling motion for all N_{Re} considered in

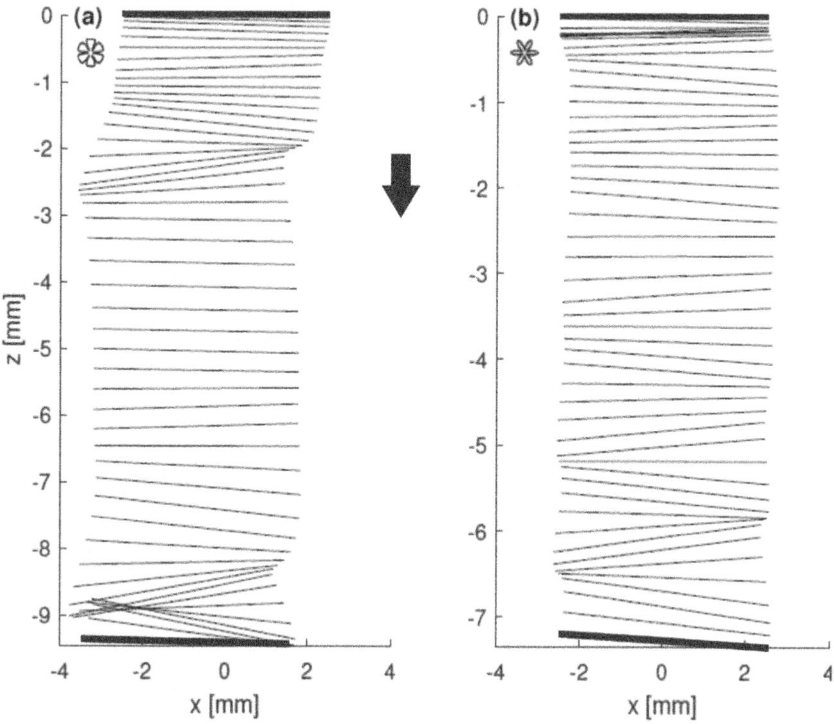

Fig. 4.30 Fall trajectories of planar ice crystals as shown through consecutive snapshots of the crystal position on the x–z plane for **a** sector plate (P1b), $N_{Re} = 384$; and **b** broad-branched crystal (P1c), $N_{Re} = 345$. The time interval is 0.005 s. The initial and final positions of the crystal are marked with thick solid bars in the projection (adapted from [35])

this study. For comparison, the highest N_{Re} for stable disk motion in the Willmarth et al. [58] tank study was 172. In that study, tumbling motions only occurred for N_{Re} in excess of 2000, much larger N_{Re} than those observed in the current study. At this moment we are unsure what causes this discrepancy because there are many differences between the dimension, shape, aspect ratio and density of the crystals in this study and Willmarth et al.'s disks. To determine the discrepancy requires further studies.

4.3.4 *Vorticity*

The structure of the wake and downstream flow, illustrated here using vorticity, may influence the collection of cloud droplets and scavenging of aerosol particles by ice crystals. The vorticity magnitude distributions on the cross section through the center of falling 5 mm sector plate (P1b, $N_{\mathrm{Re}} = 384$) and ordinary dendritic crystal (*P1e*, $N_{\mathrm{Re}} = 173$) are shown in Fig. 4.31.

As expected, the maximum in the vorticity magnitude is observed at the edges of the crystal, where the flow is changing direction and speed rapidly. The vorticity is transported downstream by the flow, with higher vorticity forming a champagne glass-shape, with the center void of high vorticity for the sector plate. The vorticity is advected downstream for the dendrite as well, not only at the outer edges, but at the interior edges as well. The vorticity is advected further downstream in the 5 mm sector plate wake than the dendrite, due to the difference in the speed of the flow—terminal velocity of 1.08 and 0.49 m s^{-1} for the plate and dendrite, respectively. The interior of the wake in the sector plate contains a region of low vorticity (less than

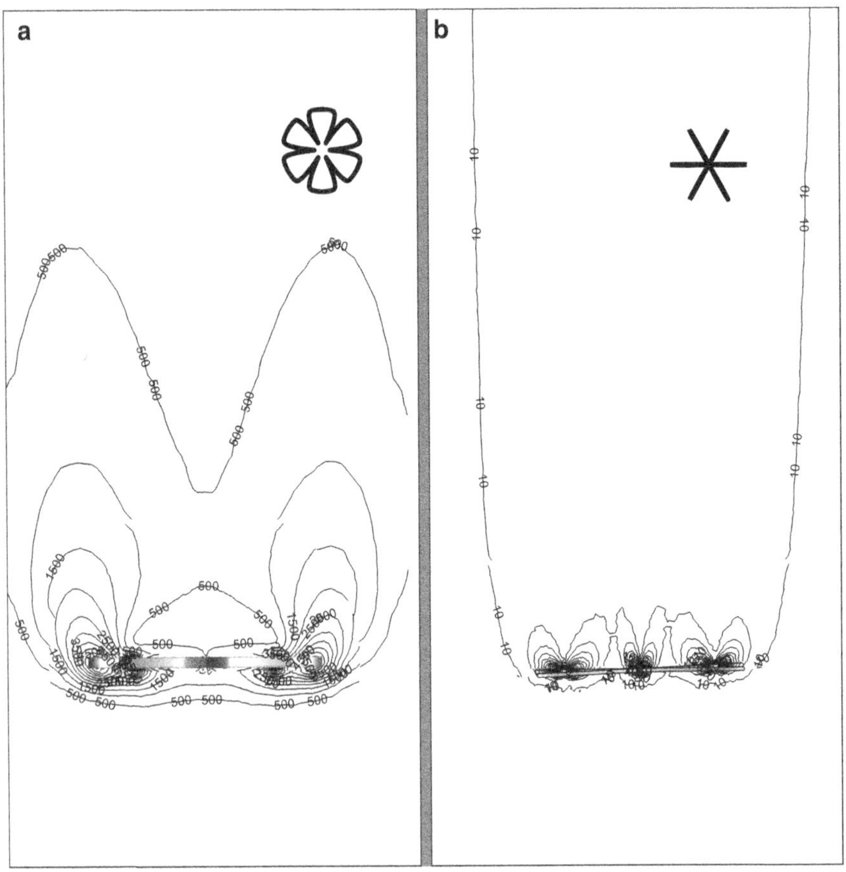

Fig. 4.31 Vorticity distribution (s^{-1}) around (**a**) an ordinary dendritic crystal *P1e* and (**b**) a sector plate *P1b*. Vorticity magnitude is shown by color shades (yellow: low; bright blue: high) over a range of 100–4000 s^{-1}. The snapshots are from randomly selected time steps (after [35])

$100\ s^{-1}$) located slightly above the surface of the crystal, appearing in a random pattern.

The vorticities form intertwining vortex tubes in the downstream wake, shown in Fig. 4.32 as three-dimensional $\hat{z}$-vorticity isosurfaces of $\pm 400\ s^{-1}$.

This structure may impact the collection of small cloud particles with subsequent ice crystals downstream by chaotically redistributing the particles in space. The

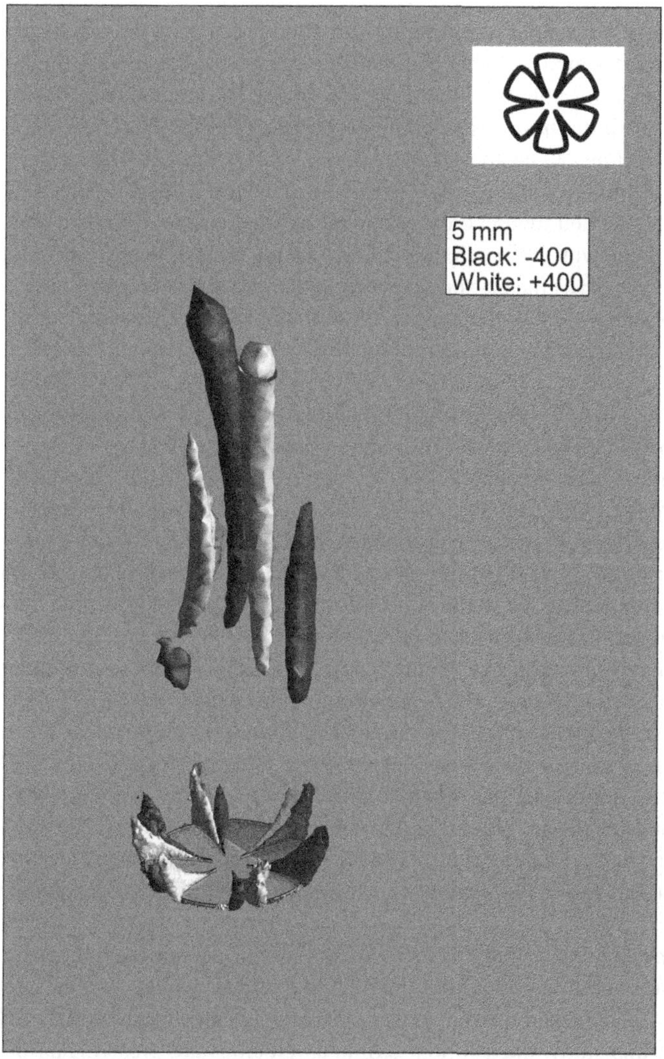

Fig. 4.32 3D $\hat{z}$-vorticity isosurface (s^{-1}) around a 5 mm sector plate (P1b) to illustrate the unsteady vortex structure. The black isosurfaces correspond to $\hat{z}$-vorticity values of $-400\ s^{-1}$ while the white isosurfaces correspond to $\hat{z}$-vorticity values of $400\ s^{-1}$. The snapshot corresponds to the same randomly selected time step, as in Fig. 4.31 (after [35])

positive and negative alternating pattern of the vorticity isosurfaces around the sector branches is intuitively understood considering the direction in which the flow will curl on either side of the branch.

4.3.5 Terminal Velocity

Similar to the hexagonal plate motions reported previously, the results of this study show that the terminal velocity of simulated ice crystals fluctuate throughout the fall process, generally bound within a certain range. The range is smaller for crystals which fall with the basal plane entirely normal to the fall direction, likely because the cross-sectional area exposed to the flow is nearly constant. The range is larger for crystals exhibiting swinging, unstable motion, because with an inclination relative to the flow, the implied smaller cross-sectional area decreases the upward drag force and increases the downward acceleration. The velocity relative to the flow initially varies from case to case as the crystal responds to the first approximation base velocity to which it is subjected. For this reason, the terminal velocity is computed by averaging the last 3000 time steps of the simulation, as per the method outlined in Cheng et al. [7].

Figure 4.33 shows the $\hat{z}$-velocity distribution on and velocity vectors around a 4 mm broad-branched crystal (P1c) and stellar crystal (P1d).

Note the scales are panel-specific, corresponding to the calculated 900 hPa terminal velocity for each crystal, 91.0 cm s^{-1} and 32.5 cm s^{-1}, respectively. Note that the velocity vectors are projections of the 3D vectors onto the y–z plane crossing the center of the plate, thus not all vectors are on the same plane, and the uneven vector distribution is due to the non-uniform mesh used to conform to the surface of the crystal and to allow for mesh adjustment with time. What we see here is that the broad-branch crystal blocks the flow more significantly than the stellar crystal due to the former's larger area, and hence the more pronounce wake low pressure region. In contrast, the stellar crystal is more skeletal and the presence of the crystal has smaller impact on the flow than does the broad branch. This is intuitively expected.

Figure 4.34 presents the 900 hPa terminal velocity for each crystal type in this study, alongside the numerical results of the 1000 hPa terminal velocity of the hexagonal plate *P1a* from Cheng et al. [7] and the 1000 hPa observational terminal velocity parameterizations of Heymsfield and Kajikawa [15] for the corresponding crystal type.

As expected, the terminal velocity increases with increasing diameter, consistent with the observational data and previous numerical results. For the sector plates (P1b), broad-branched crystals (P1c), and ordinary dendritic crystals (P1e) investigated in this study, the 900 hPa terminal velocities are greater than those of the 1000 hPa observational ones. The drag is smaller at 900 hPa level and hence it is not surprising that the terminal velocities at 900 hPa are greater than that at 1000 hPa given everything else remains the same. The only exception is the simulated stellar crystals (P1d) that have calculated 900 hPa terminal velocities consistently lower

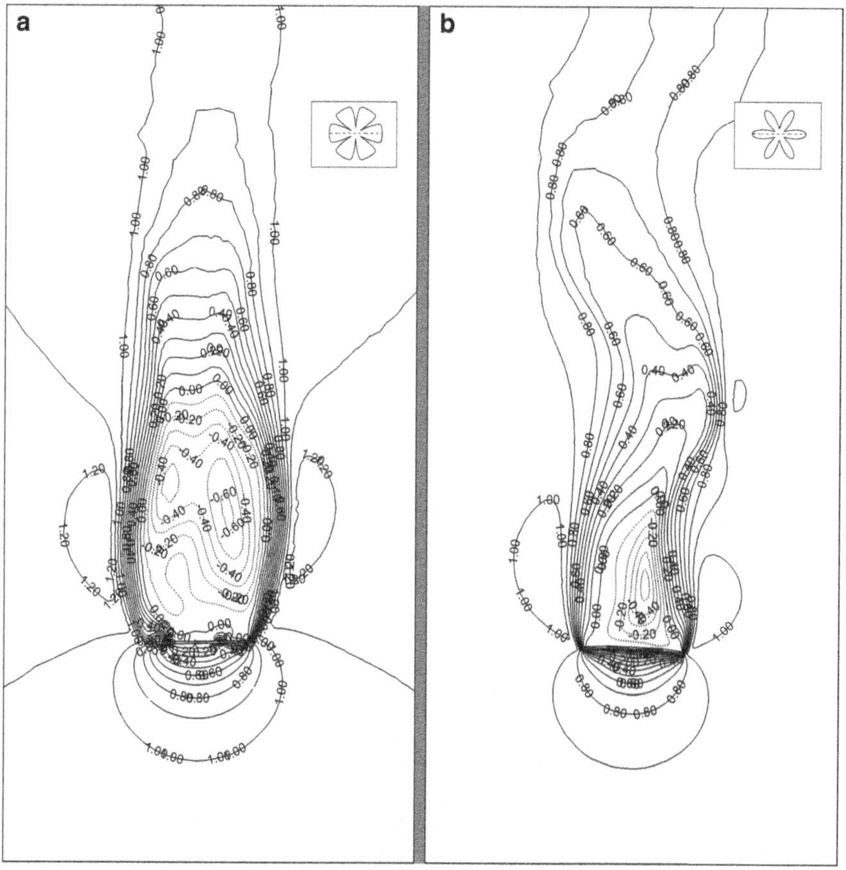

Fig. 4.33 $\hat{z}$-velocity (m s^{-1}) and velocity vector the lower pressure (2D projection of the 3D vector onto the central y–z plane) around a broad-branched $P1c$ (**a**) and stellar crystal $P1d$ (**b**). $\hat{z}$-velocity is shown by contours over a range of -0.40 to 0.40 m s^{-1}. The snapshots correspond are from randomly selected time steps (adapted from [35])

than those of the 1000 hPa observational results of Heymsfield and Kajikawa [15] for a given diameter; Heymsfield and Kajikawa [15] observations may encompass a spectrum of "stellar" shapes, whereas the calculated values in this study are from idealized shapes.

The calculated 900 hPa terminal velocities u_∞ of each crystal type can be fit by the following power law relationships with diameter:

$$\text{Sector plates } P1b\text{: } u_\infty = 9.50 \times 10^{-1} d^{0.13} - 1.76, \tag{4.12}$$

$$\text{Broad - branched crystals } P1c\text{: } u_\infty = 1.67 \times 10^{-1} d^{0.265} - 0.59, \tag{4.13}$$

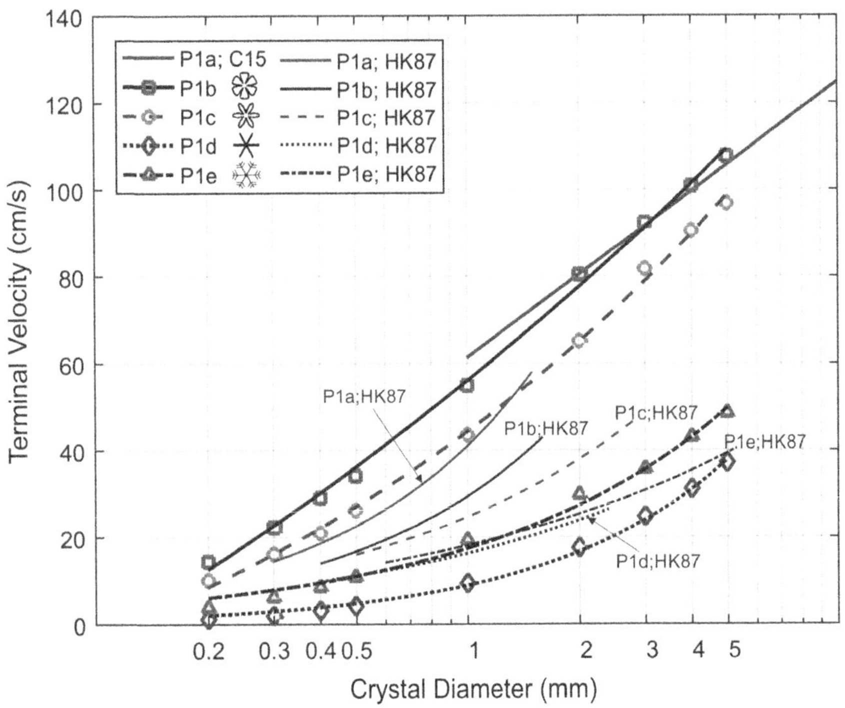

Fig. 4.34 Terminal velocities (cm s^{-1}) of planar crystals for 900 hPa. Square, circle, diamond, and triangle markers indicate the results for: sector plates (P1b), broad-branched crystals (P1c), stellar crystals (P1d), and ordinary dendritic crystals (P1e), respectively. The corresponding bold curves are the power law fits given by Eqs. (4.12)–(4.15): solid curve for sector plates, dashed curve for broad-branched crystals, dotted curve for sector plates, and dash-dot curve for dendrites. Thick, grey, solid curve represents the 1000 hPa parameterization of hexagonal plates (P1a) from Cheng et al. [7]. The corresponding 1000 hPa terminal velocity parameterizations of Heymsfield and Kajikawa [15] are shown by thin curves, with colors and line styles matching these experimental results (adapted from [35])

$$\text{Stellar crystals } P1d\text{: } u_{\infty} = 1.87 \times 10^{-4} d^{0.89}, \tag{4.14}$$

$$\text{Ordinary Dendritic crystals } P1e\text{: } u_{\infty} = 1.94 \times 10^{-3} d^{0.65}, \tag{4.15}$$

where u_{∞} has units of m s^{-1} and d is in μm. The root-mean-square error of the fits are 0.020, 0.017, 0.008, and 0.017 m s^{-1}, respectively. The relationships are valid over the range of diameters studied, 200–5000 μm for 900 hPa. Relationships between the terminal velocity and crystal surface area sa are as follows:

$$\text{Sector plates } P1b\text{: } u_{\infty} = 19.44 sa^{0.26}, \tag{4.16}$$

$$\text{Broad - branched crystals } P1c\text{: } u_\infty = 29.59 sa^{0.29}, \tag{4.17}$$

$$\text{Stellar crystals } P1d\text{: } u_\infty = 160.80 sa^{0.44}, \tag{4.18}$$

$$\text{Ordinary Dendritic crystals } P1e\text{: } u_\infty = 35.05 sa^{0.34}, \tag{4.19}$$

where u_∞ has units of m s^{-1} and *sa* is the crystal surface area in m^2. The root-mean-square error of the fits are 0.060, 0.047, 0.008, and 0.017 m s^{-1}, respectively. The relationships are valid over the range of diameters studied, 200–5000 μm for 900 hPa.

4.3.6 Drag Coefficients

Figure 4.35 shows the time-averaged drag coefficient (C_D) as a function of Reynolds number, where the drag has been computed as the average C_D over the last 3000 time steps, as described above. Drag coefficient data from other studies is presented in the figure for comparison, including numerical results for hexagonal plates *P1a* from Cheng et al. [7], Hashino et al. [14], Wang and Ji [54], and for broad-branched crystals *P1c*, also from Wang and Ji [54].

The drag coefficient for the crystal types in this study can be fit by the following two term power law relationships:

$$\text{Sector plates } P1b\text{: } C_D = 23.80 N_{\mathrm{Re}}^{-0.72} + 0.74, \tag{4.20}$$

$$\text{Broad-branched crystals } P1c\text{: } C_D = 38.85 N_{\mathrm{Re}}^{-0.87} + 1.09, \tag{4.21}$$

$$\text{Stellar crystals } P1d\text{: } C_D = 686 N_{\mathrm{Re}}^{-1.17} + 5.84, \tag{4.22}$$

$$\text{Ordinary Dendritic crystals } P1e\text{: } C_D = 123 N_{\mathrm{Re}}^{-0.72} + 0.93, \tag{4.23}$$

where all variables are dimensionless. The root-mean-square error of the fits are 0.404, 0.193, 179.9, and 1.884, respectively, with adjusted r^2 values of 0.9924, 0.9996, 0.9686, and 0.9991, respectively. Power law relationships were selected, because polynomial fits might over-fit the data and are not representative of a realistic relationship. The relationships are valid over the range of *Re* corresponding to each crystal type, found in Table 4.4.

The drag coefficients from the studies of Böhm [4] and Heymsfield and Westbrook [17] are also shown in Fig. 4.35. All their C_d are lower than that of the idealized crystals studied here. The discrepancy is most likely due to the difference in idealized versus observed crystal shapes.

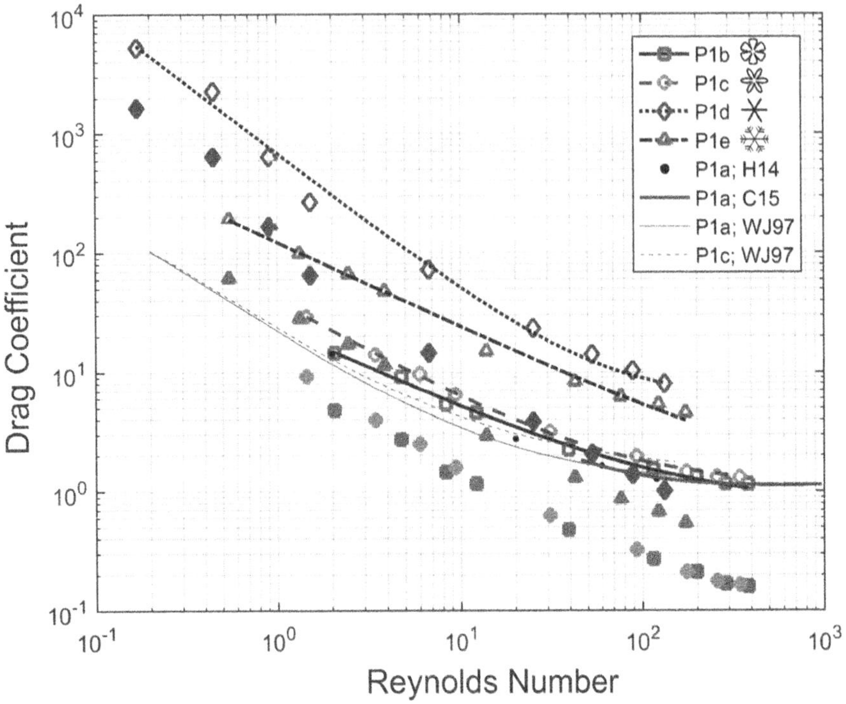

Fig. 4.35 Drag coefficient of planar crystals. Square, circle, diamond, and triangle markers indicate the results for: sector plates (P1b), broad-branched crystals (P1c), stellar crystal (P1d), and ordinary dendritic crystals (P1e), respectively. The corresponding curves are the power law fits given by Eqs. (4.20)–(4.23): solid curve for sector plates, dashed curve for broad-branched crystals, dotted curve for sector plates, and dash-dot curve for dendrites. Bolded bullets represent the steady state drag coefficients for hexagonal plates (P1a) from Hashino et al. [14]. Thick, grey, solid curve represents the parameterization of hexagonal plates from Cheng et al. [7]. The parameterizations of Wang and Ji [54] are shown in thin lines: solid curve for hexagonal plates, dashed curve for broad-branched crystals. The adjusted Cd discussed in Heymsfield and Westbrook [17] ($C_D^* = C_D A_r^{0.5}$) are shown as the filled in markers, shades and shapes corresponding to the crystal type described in legend (adapted from [35])

4.4 Fall Behavior of Snow Aggregates

As mentioned in Chap. 2, what commonly called snowflakes are actually aggregates of ice crystals formed by single ice crystals clumping together upon collision. The number of ice crystals involved can be from several to dozens depending on the size of the flake. When the number of ice crstals increases, the shape quickly becomes complicated. For the purpose of numerial calculations of flow fields, it is very expensive to perform simulation of cases with high number of crystals. We are currentlt performing calculations of the flow fields around freely falling snowflakes using aggregates from 2 to 6 dendrites to at least see a trend so that we can extrapolate from these results to those of higher number aggregates.

In the following, we shall use the preliminary results of the case of a 3-dendrite aggregate to illustrate the fall behavior of snowflakes. The first step of peforming the simulation is, as usual, defining the shape and size of the particle. Due the lack of systematic study of aggregates so far, we can oly form these particles using some aggregate photographs as the base. We formed the aggregate by manually attach 3 dendritic crystals of different sizes together in a manner more-or-less similar to what we see in a snowflake photograph. Figure 4.36 shows this 3-dendrite aggregate as viewed from different angles.

Figure 4.36 shows that an aggregate is a highly 3-diemsional object and very asymmetric, and the numerical meshing becomes highly complicated because of the many small spaces between the branches. Computational crashes are often because of the instability caused by very thin meshes at certain places. Many trial and error tests were performed before stable solutions are obstained.

Figure 4.37 shows the sequence of the aggregate position for time steps 1050, 1500, 2000, 2500, …, 5000. In the first 3 steps, the aggregate does not translate too

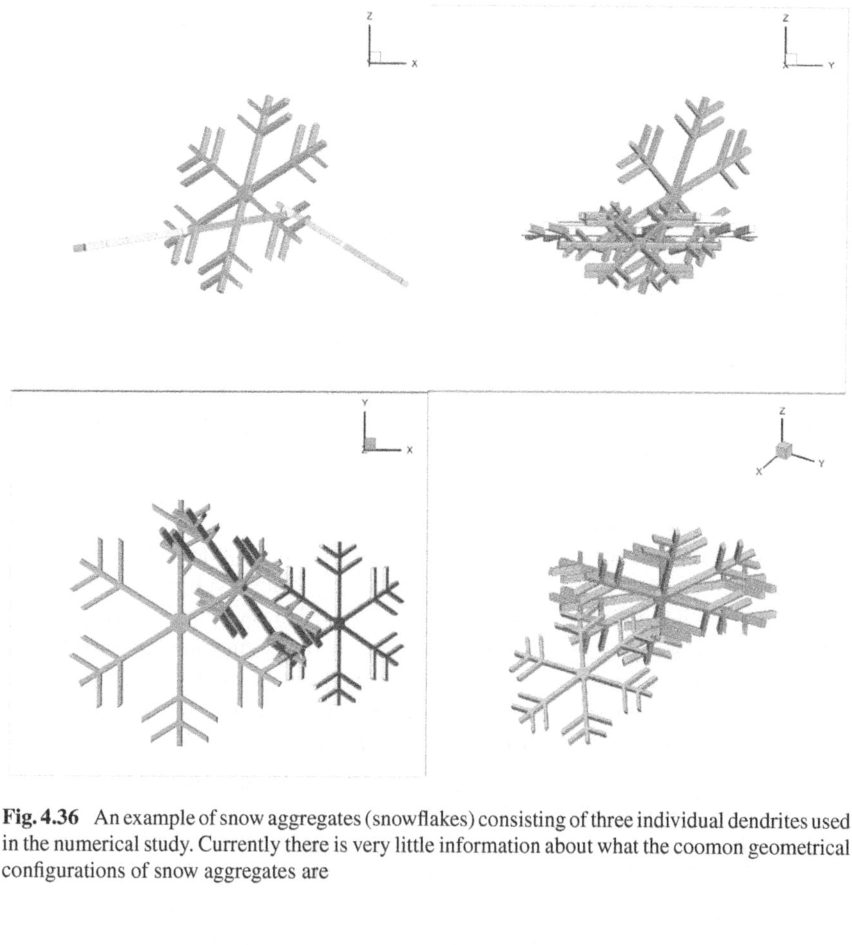

Fig. 4.36 An example of snow aggregates (snowflakes) consisting of three individual dendrites used in the numerical study. Currently there is very little information about what the coomon geometrical configurations of snow aggregates are

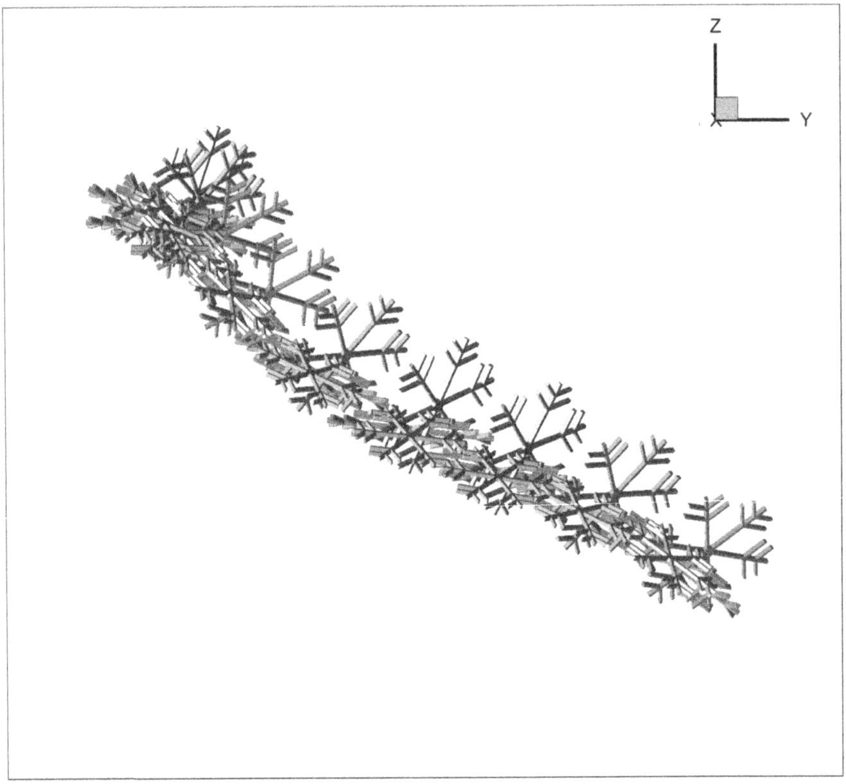

Fig. 4.37 The sequence of the aggregate position (from upper left to lower right) for time steps 1050, 1500, 2000, 2500, …, 5000

much but makes a semi-spiral path while descending. After that, it makes a relatively smooth translation to the side albeit with a mild wavy motion in the vertical direction. The more dramatic motion in the beginning appears just as an adjusting stage so that the aggregate respond to a torque that is greater at the beginning. As time goes on, the torque decreases and thus the aggregate performs smooth translation without much twists and turns. This seems to conform with the general experience of observing a small aggregate floating in a calm air during the beginning of a light snow event.

Figure 4.38 shows the xz-view of the pressure distribution and streamtrace pattern around the falling aggregate at t = 0.25 s. From a distance, the pressure distribution is not too different from that of a plate, namely, the front side is high pressure region and the rear is a low pressure region. However, due to the complicated structure of the aggregate, the pressure distribution near the surface is understandably rather complex, but it is important for the understanding of the aggregation (or clumpin) process of ice crystals to form snowflakes.

Some streamtraces are also shown in Fig. 4.38. While the motion of the aggregate is unsteady, the instantanous flow field is not too complex as the Reynolds number is

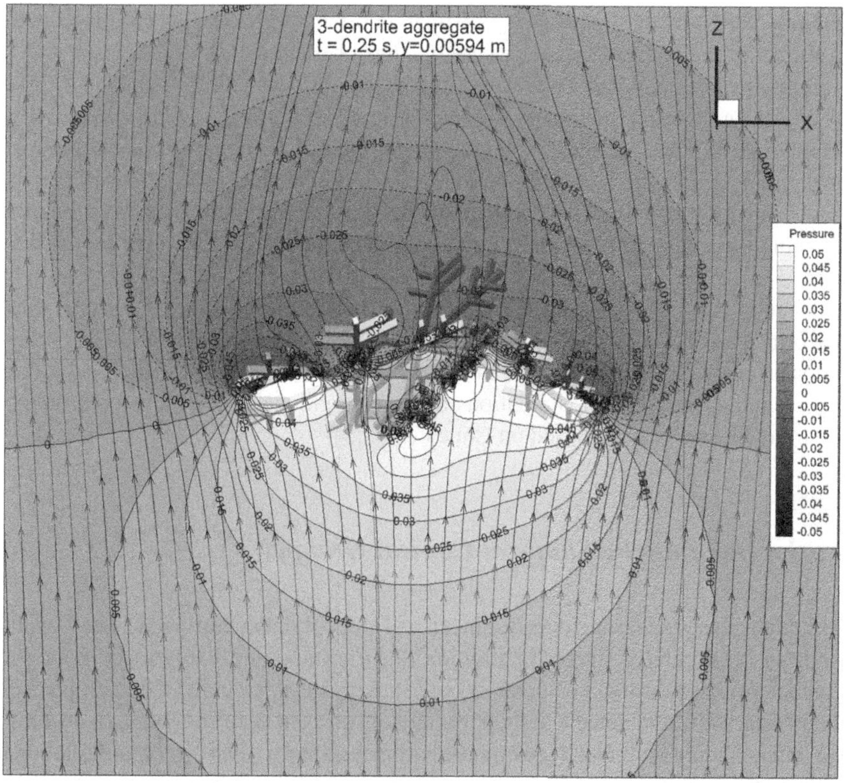

Fig. 4.38 The xz-view of the pressure distribution and streamtrace pattern around the falling aggregate at t = 0.25 s (adapted from [35])

relatively small due to the small fall velocity. The front side is mostly laminar while the wake region has a few loops. At present, onlt preliminary results are obtained and further results will be reported in the near future.

4.5 Motion of Falling Conical Graupel

As mentioned in Chap. 2, graupel (or soft hails) are the precursor of hailstones and often conical in shape. In cloud physics, graupel is defines as an ice particle grown mainly by riming process and with a size less than 5 mm [41, 57]. Further growth of graupel will lead to the formation of hailstones.

The motion of graupel has not been studied theoretically until Kubicek and Wang [27] performed the first successful numerical simulation of flow around a falling conical graupel of a diameter 3 mm. At the time, the fall of the graupel was not quite “free” due to the limitation of computing resource so that the graupel was assume to fall only vertically but with varying inclination orientations. The purpose is to study the effect of the inclined orientation on the flow characteristics.

Subsequent to Kubicek and Wang [27], Wang and Kubicek [55] studied a series of conical graupel of diameters 0.5–5 mm that fall vertically with fixed vertical orientation. While both studies shed lights on the fall behavior of conical graupel, they are limited by the constraint of no feedback of flow field to the graupel, i.e., the studies are basically that of the flow past a fixed obstacle of conical shape instead of allowing the graupel to respond to the hydrodynamic forces generated by the flow field.

Chueh et al. [9] performed a new study that removes this deficiency. They performed a series of simulations that allow the graupel velocity, position and orientation to respond to the changes in the flow field as time goes. The results thus represent the true free fall simulations for conical graupel. The results presented in the following section are based on Chueh et al. [9].

4.5.1 Defining the Shape of Conical Graupel

Although most graupel are indeed conical, real graupel shape can deviate significantly from an ideal conical shape. In addition, their surfaces are likely to be grainy as graupel grow via riming, i.e., collision and sticking of supercooled cloud droplets on their surfaces. But to make a first approximation so as to keep computation cost low, we will deal only with idealized conical particles with smooth surface and rotational symmetry about the z-axis.

The shape and size of the conical graupel is based on the equation depicting the x–z-cross-section of the graupel shape proposed by Wang [49]:

$$x = \pm a\sqrt{1 - \left(\frac{z}{c}\right)^2} \cos^{-1}\left(\frac{z}{\lambda c}\right), \tag{4.24}$$

where x and z are the horizontal and vertical coordinates whereas a and c are used to control the horizontal and vertical semi-axis lengths, respectively. The parameter λ, whose value varies between 1 and ∞, serves to control the sharpness of the apex of the conical graupel. For more details about this expression and the shape it generates, see Wang [49].

Equation (4.24) generates only the xz-cross-section of the graupel. To obtain a 3D conical body, we revolve this cross-section about the z-axis which yields the following 3D mathematical expression:

$$\frac{x^2}{\left(a\ \cos^{-1}(z/\lambda c)^2\right)} + \frac{y^2}{\left(a\ \cos^{-1}(z/\lambda c)^2\right)} + \frac{z^2}{c^2} = 1, \tag{4.25}$$

which is a special case of the conical bodies with elliptical cross-sections descried by Wang [52]. Figure 4.39 shows the shape generated by (4.25).

Using Eq. (4.25) and assuming various combinations of a, c and λ, we can generate conical graupel of different sizes and shapes [49]. The parametrical combinations we use in the present study are the same as that in Chueh et al. [8]. The boundary conditions are the same as described in Eqs. (3.5)–(3.8).

In the following, the flow characteristics around freely falling conical graupel particles of diameter 0.5, 1, 2, 3, 4, 5 mm will be described. As mentioned before, graupel becomes hailstones when they are larger than 5 mm. Here we assume that the pressure is 800 hPa and temperature is −8 °C. In such an environment, the temperature is in the warm range (from 248 to 278 K) where graupel particles are typically found [41]. The atmospheric environment is the same as those used previously in Wang and Kubicek [55], Hashino et al. [13, 14], and Chueh et al. [8]. The graupel's far stream velocity U_3 values were set to the terminal fall velocities estimated based on the empirical velocity-mass relationship for conical graupel by Locatelli and Hobbs [29]. The values are calculated based on a mathematical equation in their Table 1 of [29].

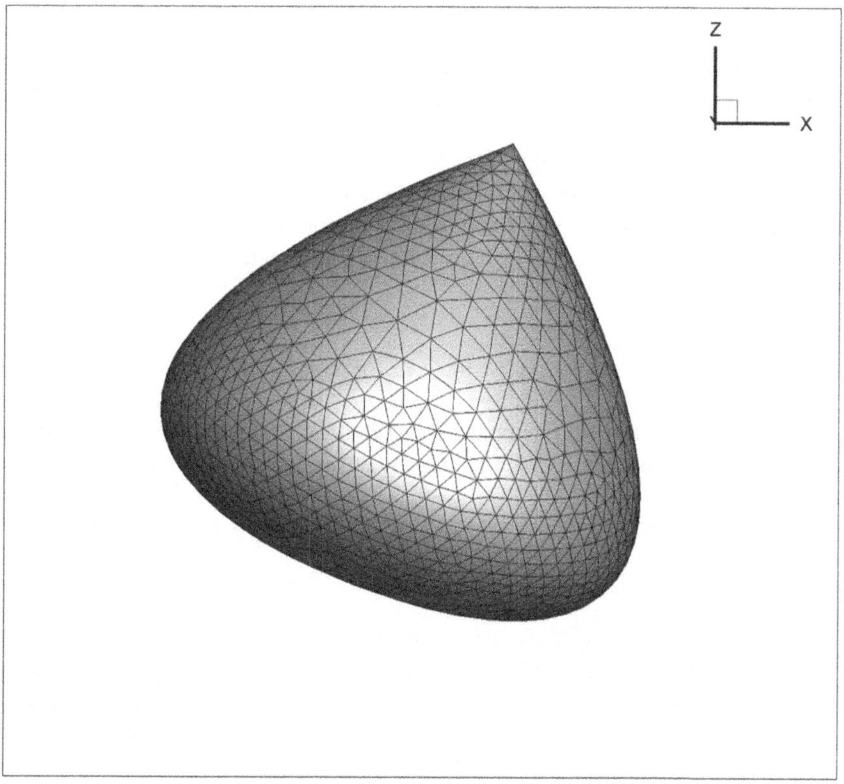

Fig. 4.39 The shape of conical graupel generated by Eq. (4.25). Shown on the surface is the numerical mesh generated by FLUENT

We produce an initial perturbation in the flow field around the graupel to quickly initiate the unsteady motion of the falling graupel. Similar to that done for other hydrometeors described previously, this is done by using an initial inclination angle of 20° counterclockwise around the y'' axis. The small initial inclination angle is intended to produce a correspondingly small perturbation, which induces rotating oscillations of the falling graupel. Then, a steady flow solution is obtained for the inclined state of the graupel with a constant U_3 and serves as the initial condition of the freely falling conical graupel simulation.

4.5.2 Dimensions of Conical Graupel

The shape parameters a, c and λ of the graupel as well as their corresponding far field velocities, and the terminal velocities of the graupel are given in Table 4.6.

The negative V_t values indicate that graupel are falling relative to the computational domain which is the computational "wind tunnel". We divide the wind tunnel into a few different computational zones similar to that used by Hashino et al. [13] and Chueh et al. [8]. The mesh of one of the computational zones, the one closest to the outer boundary of the entire domain, is always kept unchanged throughout every of the simulations performed here, whereas the meshes of the rest zones either change or move with the graupel at every time step, as the falling graupel changes its orientation and position with time.

Because the motion of the graupel is unsteady and its velocity can change substantially with changing fall attitude, there is a need to define two different Reynolds numbers so as to describe the instantaneous motion in better clarity. We use the following two definitions of Reynolds number. The regular Reynolds number is defined as

$$N_{\mathrm{Re}} = \frac{du_{\infty}}{\nu}, \tag{4.26}$$

where d is the diameter of the graupel representing its maximum width in x and y directions, and u_{∞} is equal to U_3. The quasi-steady state Reynolds number is defined as

$$N_{\mathrm{Re}q} = \frac{d(u_{\infty} - V_t)}{\nu}, \tag{4.27}$$

where the relative terminal fall velocity values of V_t are obtained from the simulations that already reach their individual steady-state fall velocity values (Vel_z). The quasi-steady state Reynolds numbers are considered as the actual ones that are calculated and listed in Table 4.5.

Table 4.5 Geometric and kinematic characteristics of the graupel investigated in this study. See the text for detailed definitions

d (mm)	a (mm)	c (mm)	u_∞ (m s^{-1})	V_t (m s^{-1})	$N_{Re\ (normal)}$	$N_{Req\ (quasi\text{-}steady)}$	Graupel density (g cm^{-3})
0.5	0.137	0.229	0.5	−0.17	15.6	21	0.245
1	0.275	0.458	0.97	−0.1754	60.7	71.6	0.186
2	0.55	0.916	1.67	−0.27	209	243	0.141
3	0.824	1.374	2.06	−0.3	387	443	0.12
4	1.099	1.832	2.32	−0.34	580	665	0.107
5	1.373	2.290	2.55	−0.48	797	948	0.0977

The densities of the graupel investigated here are also listed in Table 4.5. The density was determined by dividing the graupel mass observed by Locatelli and Hobbs [29] by the volume calculated from Eq. (4.25). These values are much smaller than the constant density 0.91668 g/cm^3 of solid ice used in Chueh et al. [8] but are more realistic, as most real graupel particles are mixtures of solid and porous ice particles, and the density should be smaller than that of solid ice.

4.5.3 *Fall Attitudes and Flow Characteristics*

The simulation results show that the fall attitudes of this group of graupel can be classified into 3 categories: (1) damped oscillation ($d = 0.5$ mm); (2) persistent oscillation but no tumbling ($d = 1$ and 2 mm); (3) persistent tumbling ($d \geq 3$ mm). These will be described in the following.

4.5.3.1 Damped Oscillation ($d = 0.5$ mm)

As mentioned before, we produced a small perturbation in the flow field by tilting the graupel vertical axis 20°. The viscous effect of air soon diminishes this small perturbation and the graupel returns to upright orientation and falls vertically.

Figure 4.40 shows the time behavior of a few key kinematic variables of the fall motion of a graupel of $d = 0.5$ mm.

The damped oscillation phenomenon, e.g., cg_x in (a), Vel$_x$ in (c), θ_y in (e) and T$_y$ in (f), is obvious. They all demonstrate that the perturbation caused oscillation is damped by the viscous effect of air. The rest of variables, such as cg$_y$, cg$_z$, Vel$_y$ and Vel$_z$ exhibit little or non-oscillatory behavior, as they are expected to be. In addition to oscillation, Fig. 4.45a also shows that the x–y position of the graupel has moved away from the origin, especially in the x-direction. The y-position changed little.

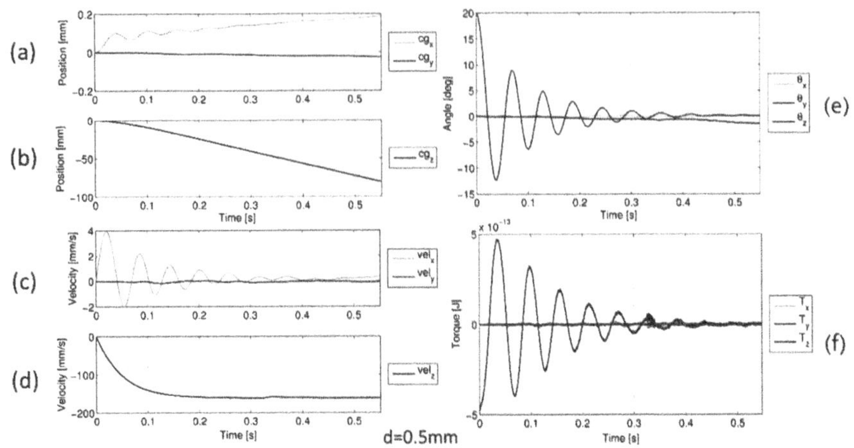

Fig. 4.40 Time variation of a few key kinematic variables for a falling conical graupel of d = 0.5 mm: **a** the x and y positions of COG, cg_x and cg_y, respectively; **b** the z position of COG, cg_z; **c** x and y-components of the graupel velocity vel_x and vel_y; **d** vel_z; **e** Tait–Bryan angles θ_x, θ_y and θ_z; **f** torques T_x, T_y, T_z (adapted from [9])

Figure 4.41 shows the plot of torque acting on the graupel as a function of the tilt angle. The starting tilt angle is 20° (the blue circle on the extreme right).

We see that the torque increases initially until the tilt reaches the negative maximum as the graupel oscillates, then decreases again as the graupel swings back.

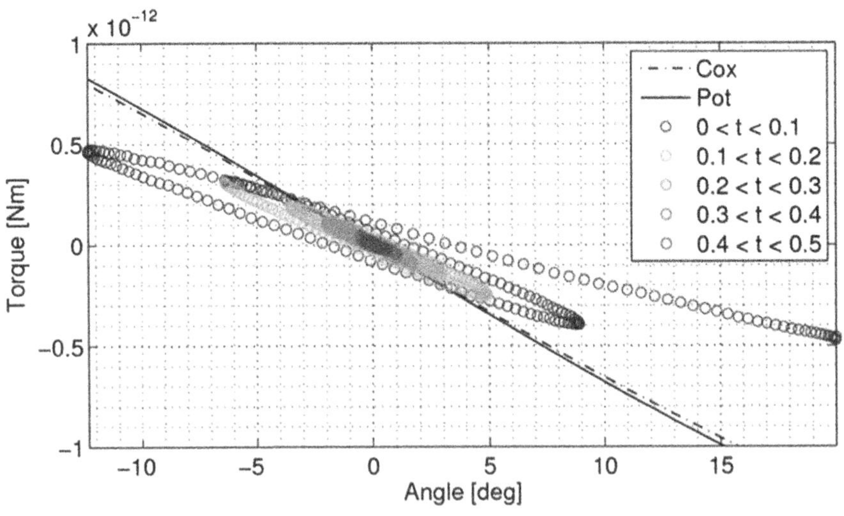

Fig. 4.41 The torque on the graupel of d = 0.5 mm as a function of θ_y. Different colors represent different time periods (in seconds) as indicated in the legend box (adapted from [9])

Eventually the torque settles in the purple and red region in Fig. 4.41 as the oscillation is dwindling. We see that the final stage of the torque variation with angle is consistent with that predicted by the theory of Cox [11] and that of potential flow for small spheroids (see [13], for more details).

As indicated above, the flow field soon reaches steady state after an initial perturbation. Figure 4.42 shows, from left to right, the pressure and z-speed distributions in the central x–z plane, and the streamtraces pattern of the steady state flow field of a conical graupel of d = 0.5 mm falling at terminal velocity. Flow fields similar to this have been described in detail by Kubicek and Wang [27] and Wang and Kubicek [55].

The left panel shows the pressure distribution. Below the graupel base is the high pressure region as it should be. The low pressure region is a quasi donut-shaped ring around the "equator" of the falling graupel. The middle panel shows the z-velocity distribution. The low velocity region is around the graupel and immediate to the surface is an envelope of negative (downward) z-velocity region. The high velocity region is an outer donut-shaped region surrounding the equatorial low velocity region.

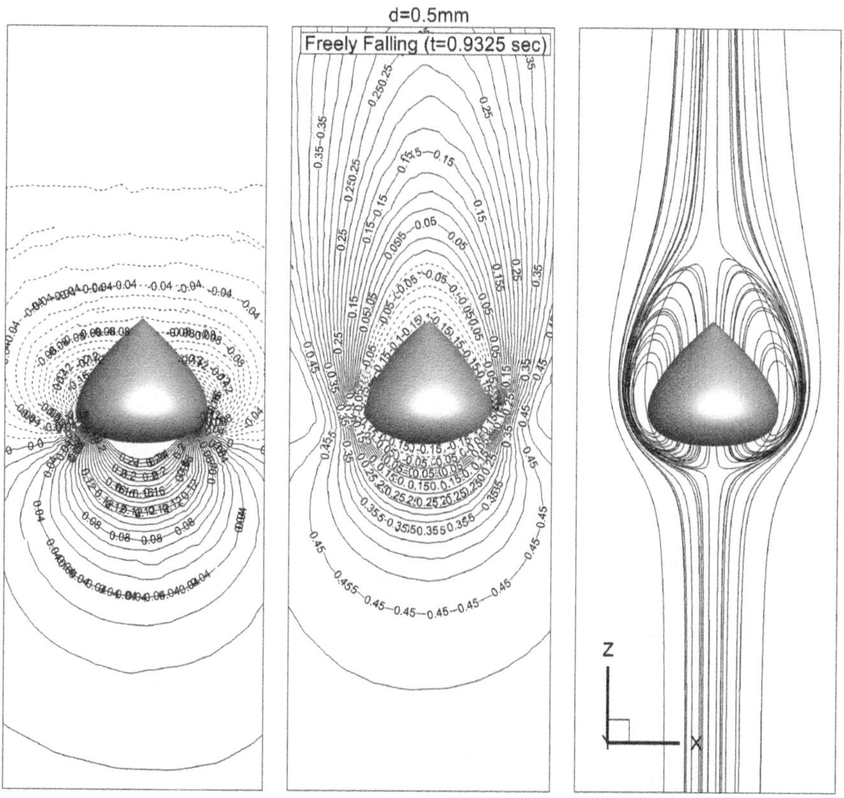

Fig. 4.42 A snapshot of the flow field in the central xz-plane around a falling conical graupel of $d = 0.5$ mm: (left) pressure distribution, (center) z-velocity distribution, and (right) streamtraces

The right panel of Fig. 4.42 shows the streamtrace pattern. It is clearly seen that the forward region has a stagnation point. Interestingly, the pattern around the graupel looks similar to that "inside" a conical liquid drop falling in air with a left-right symmetric vortex ring.

4.5.3.2 Persistent Oscillation But No Tumbling ($d = 1$ and 2 mm)

For larger conical graupel of $d = 1$ and 2 mm, the fall behavior turns to that of persistent and periodic oscillation but no tumbling. We shall use the case of $d = 1$ mm to illustrate this fall attitude.

Figure 4.43 shows a group of plots of kinematic variables as a function of time.

Now the persistent periodic motion is most obvious in cg_x, Vel_x, θ_y and T_y. At the same time, we also see that there is also a corresponding periodic motion in y-direction as demonstrated by the curves of cg_y, Vel_y, θ_x and T_x, although the amplitudes are smaller. The oscillation in the y-direction exhibits an amplifying tendency. The same happens in the corresponding kinematic variables in the z-direction but the amplitudes are even smaller. Obviously the zigzag motion in the x-direction must have influenced the flow field such that the graupel motions in the y- and z-direction also become periodic.

We did not investigate the possibility that the amplifying oscillation in the x- and z-direction would eventually lead to tumbling, as this requires to run the calculation for much longer time. We will investigate this possibility in the future.

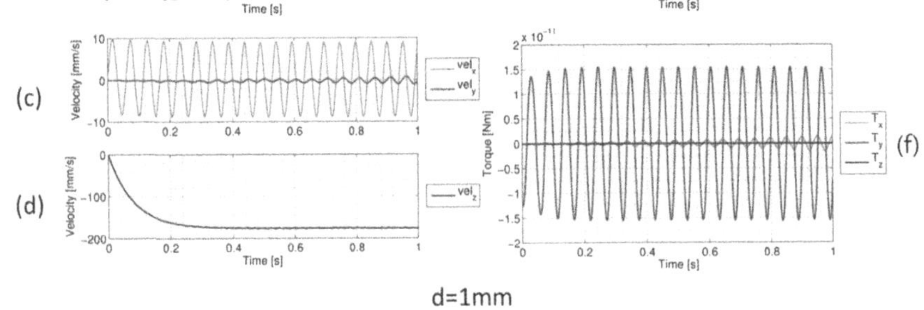

Fig. 4.43 Time variation of a few key kinematic variables for a falling conical graupel of $d = 1$ mm: **a** the x and y positions of COG, cg_x and cg_y, respectively; **b** the z position of COG, cg_z; **c** x and y-components of the graupel velocity vel_x and vel_y; **d** vel_z; **e** Tait–Bryan angles θ_x, θ_y and θ_z; **f** torques T_x, T_y, T_z (adapted from [9])

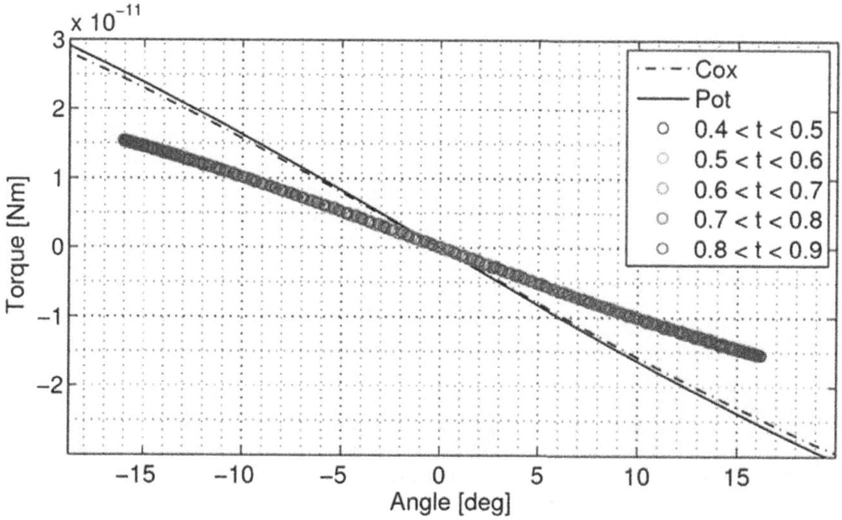

Fig. 4.44 The torque on the graupel of $d = 1$ mm as a function of θ_y. Different colors represent different time periods (in seconds) as indicated in the legend box (adapted from [9])

From Fig. 4.43a, it also appears that cg_y is increasingly negative, i.e., the y-position of the graupel deviates more and more from its original place while cg_x remains oscillatory at about the same x-coordinate. Overall, this indicates that, as time goes on, the graupel will translate horizontally further and further away from its original position. This fact may be of some importance to cloud physics and we will expand more on this point later.

Figure 4.44 shows the calculated torque as a function of tilt angle. We see that the relation is now quite different from that predicted by Cox [11] and the potential flow theory. This is not surprising as Cox's theory only applies to very small Reynolds numbers while the potential flow theory ignores the viscous effect totally. Both are not true for the present case.

Figure 4.45 shows, from left to right, the pressure, z-velocity and streamtrace fields, respectively, of a randomly chosen snapshot of the unsteady flow field around the falling graupel of $d = 1$ mm. Unlike the $d = 0.5$ mm case, this graupel now performs persistent periodic oscillatory motions in the horizontal plane during its fall. The magnitudes of the pressure deviations and the size of the recirculation bubble are larger than the $d = 0.5$ mm case, as they should be.

Note that the present unsteady flow case is different from the cases reported in Wang and Kubicek [55] where the graupel is held at a fixed orientation, hence no flow field feedback on the motion of the graupel is possible there. In the present case, the graupel is allowed to adjust its position and orientation according to the hydrodynamic force and torque acting on it, and therefore one should not expect the flow field in the present case be the same as the graupel in Wang and Kubicek [55]

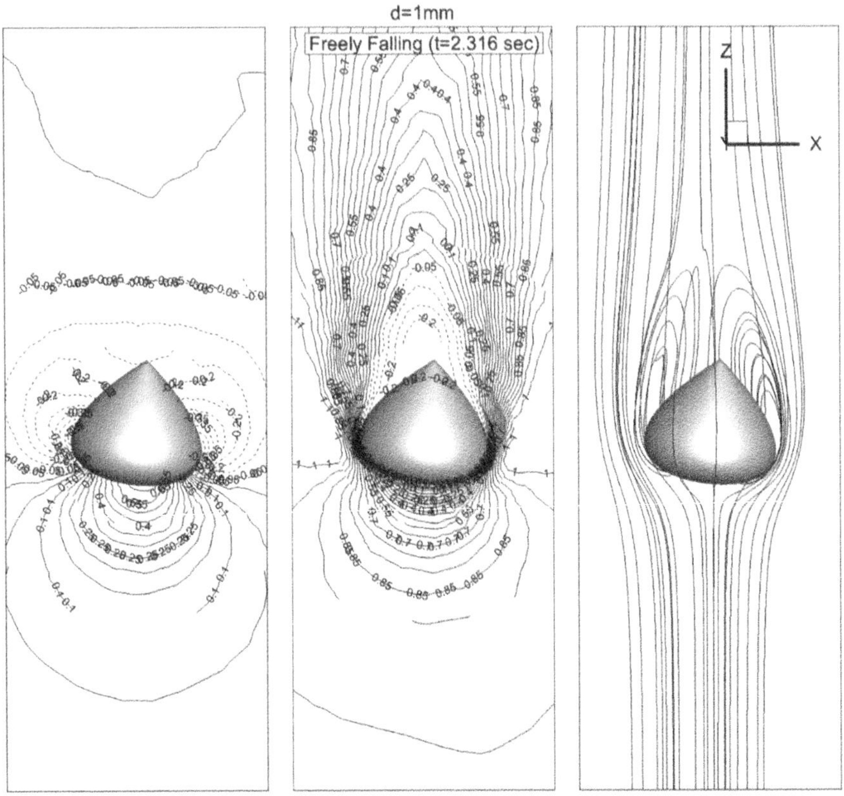

Fig. 4.45 A snapshot of the flow field in the central *xz*-plane around a falling conical graupel of $d = 1$ mm: (left) pressure distribution, (center) z-velocity distribution, and (right) streamtraces

even if the size, shape, and 3-D orientation of the two graupel are completely the same at that moment.

4.5.3.3 Persistent Tumbling ($d \geq 3$ mm)

For graupel of $d \geq 3$ mm, the fall attitude changes to persistent tumbling. We will use the $d = 5$ mm graupel to illustrate such fall attitude. Figure 4.46 shows a random frame of the pressure, z-velocity and streamtrace fields in the central xz-plane of the flow field around a falling conical graupel of $d = 5$ mm.

Figure 4.51 shows the time variation of kinematic variables of the 5 mm graupel.

We now see that the fall attitude of this graupel differs significantly from the previous two categories. Figure 4.46a shows that both cg_x and cg_y move away from the origin more and more as time goes on, implying that the graupel translates a significant distance horizontally. This sideway translation is far greater than that in the 1 mm case.

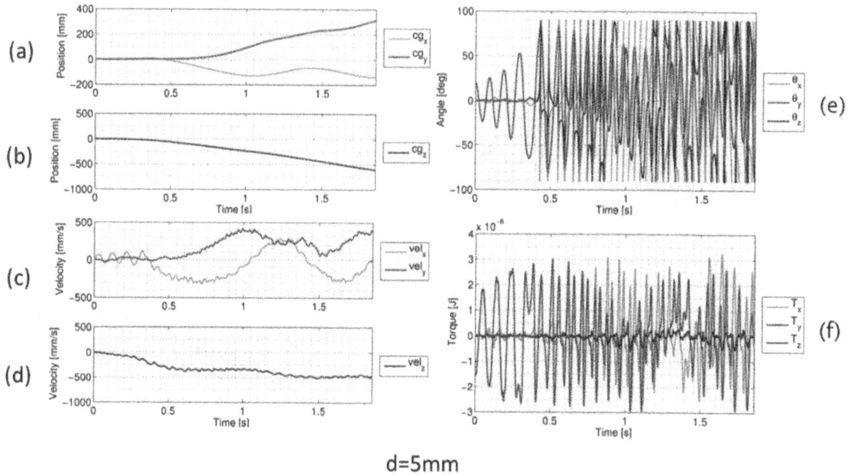

Fig. 4.46 Same as Fig. 4.43 except for $d = 5$ mm (adapted from [9])

The tumbling motion triggers complex motion modes. Figure 4.46c shows that Vel_x and Vel_y change significantly with time. This obviously is associated with the tumbling motion of the graupel. We see that the variation of Vel_x lags that of Vel_y, implying that the change in the latter induces the response of the former. There is also a more subtle response in Vel_z (Fig. 4.46d) which appears to stabilize after t ~ 1.3 s.

Figure 4.46e shows the time variation of the three Tait–Bryan angles. It can be seen that the while θ_y is oscillatory throughout the simulation, both θ_x and θ_z go off the scale after t ~ 0.4 s. The latter indicates that the graupel becomes unstable and begins to tumble at this time.

To see clearer this behavior, we replot the time variation of θ_x and θ_y using different scales as shown in Fig. 4.47.

Here we see that θ_x curve starts to go upward with values greater than 360° after t ~ 0.4 s, indicating that it tumbles over and over many cycles until t ~ 1.3 s when it begins to tumble in opposite direction.

Figure 4.48 shows the change of torque with time of the falling 5 mm graupel.

The torque behavior is much more complex than the two previous cases. Before tumbling occurs, the torque behaves more or less like that of the 1 mm graupel (blue circles), i.e., it already deviates significantly from the predictions by Cox [11] and potential flow theory. As time goes on, the difference become greater (cyan, green and purple circles) and the nonlinear nature of the relation (the "S" shape of the curve) between the torque and angle becomes more and more pronounced. Eventually tumbling occurs after t ~ 0.4 s (red circles) and the torque stops to be a single-valued but turns into a multivalued function of the tilt angle.

Tumbling of graupel was reported in the wind tunnel experimental studies of graupel by Pflaum et al. [36]. Since graupel is the precursor of hailstones and most of hailstones are spheroidal (and thus lack the sharp apex like the conical graupel),

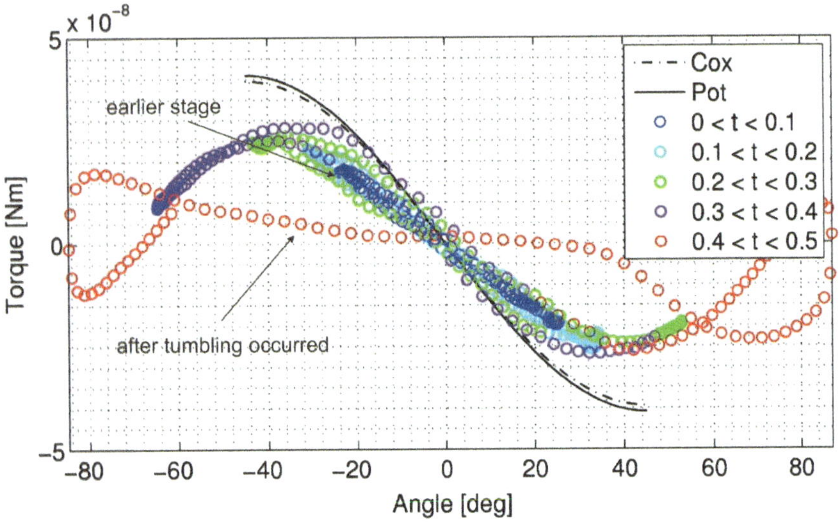

Fig. 4.47 Same as Fig. 4.44 except for $d = 5$ mm (adapted from [9] with changes)

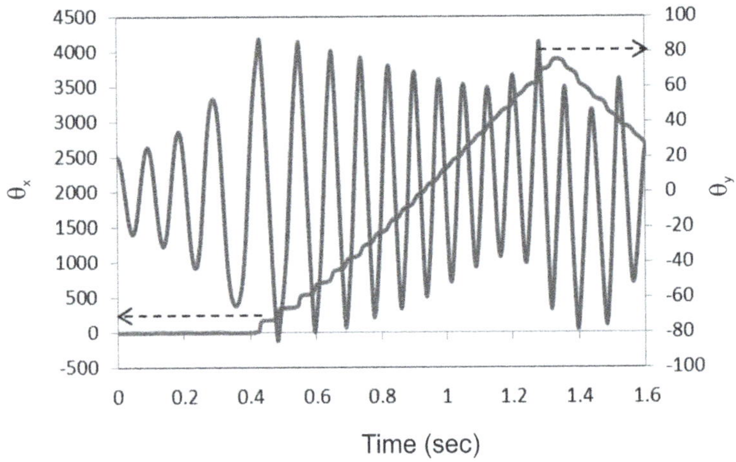

Fig. 4.48 Tait–Bryan angles θ_x, θ_y as a function of tome for $d = 5$ mm. Note there are two angle scales

we theorize that this may be the result of tumbling starting from the graupel when the latter become large enough.

Figure 4.49 shows a random frame of the pressure, z-velocity and streamtrace fields in the central xz-plane of the flow field around a falling conical graupel of $d = 5$ mm.

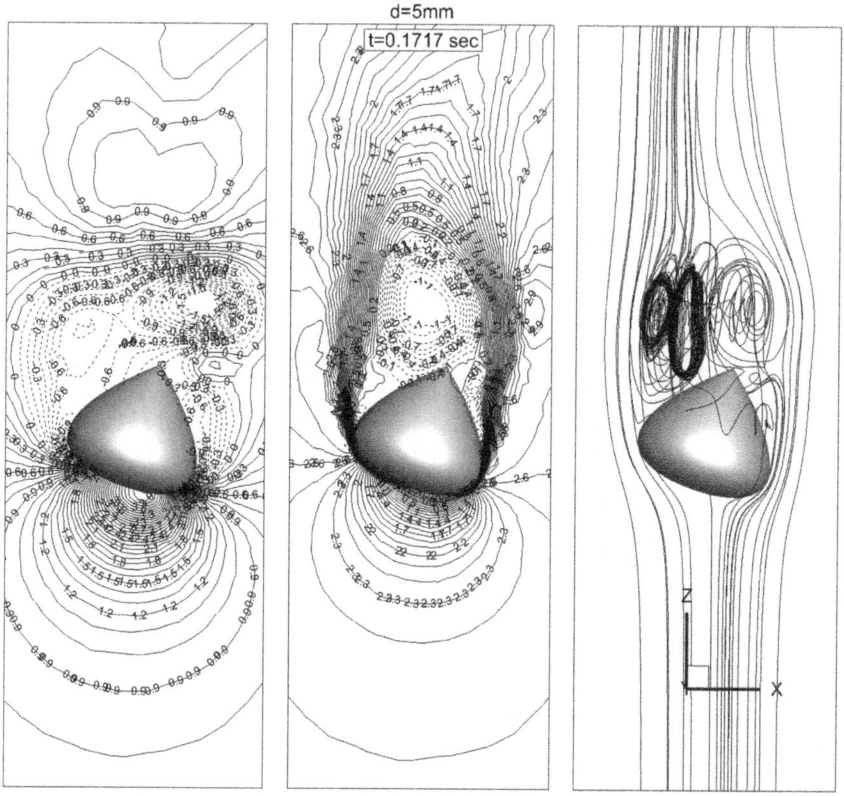

Fig. 4.49 Same as Fig. 4.42 except for d = 5 mm

In the above we have reported the kinematics of the tumbling. To understand the tumbling dynamics of conical graupel, we will need to make more careful analysis of the flow.

Note that all the results reported above in (1), (2) and (3) are based on the initial tilt $\theta_y = 20°$ which is a very modest tilt. What if the initial tilt angle is larger than 20°? We have not yet made calculations on such cases but we feel it is likely to behave more or less the same, or perhaps tumbling will occur to graupel smaller than 3 mm. This will be left for m-3 future studies are, however, very similar to the o.

In another set of numerical experiments on conical graupel performed earlier by Chueh et al. [8] where the density of graupel is assumed to be constant at 0.916 g cm^{-3} and the surrounding condition is 800 hPa and −8 °C, the same as the cases presented above. Because of the higher density, the Reynolds numbers are somewhat higher than those reported above, however, the flow characteristics are similar for similar sized graupel. As mentioned in Chap. 2, List and Schemenauer [28] of the flow observed that a cone can fall persistently with apex pointing downward. To test this possibility, Chueh et al. [8] performed a simulation with the apex pointing at an

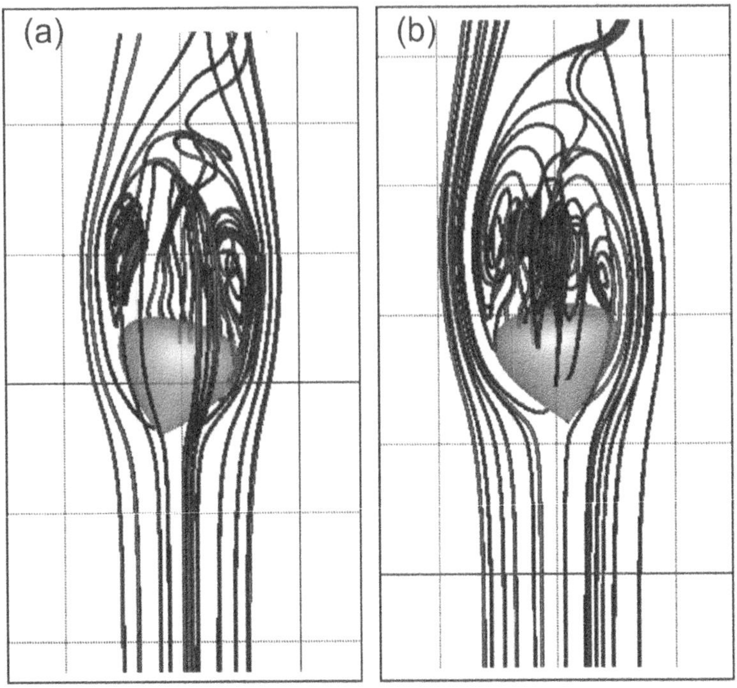

Fig. 4.50 Time evolution of the streamtrace patterns of the $d = 5$ mm conical graupel with its initial inclination of 160° with respect to Y axis: **a** $t = 0.29$ s, **b** $t = 0.5$ s (adapted from [8])

inclination angle of 160° (i.e., 20° from vertically downward). Figure 4.50 shows a few snapshots of the flow fields of such falling graupel of $d = 5$ mm.

Figure 4.51 shows the fall patterns of such downward pointing graupel of various diameters. It demonstrates that the downward pointing fall attitude can indeed be stable as observed by List and Schemenauer [28].

However, although such downward pointing fall attitude is possible, we still think that the majority of the graupel should fall with the apex pointing upward.

4.5.4 Horizontal Displacement

It is seen in the above discussions that conical graupel may perform substantial horizontal movements during the fall. Figure 4.52 shows the trajectory of the 5 mm graupel as it falls from its original position (0, 0, 0) during t = 0 to t = 1.68 s.

Figure 4.52a shows that the trajectory is a complicated spirally curve that leads to a horizontal position that is about 0.35 m away from the origin. Figure 4.52b shows that the graupel orientation and horizontal position keeps changing as it falls. The simulated fall attitudes include all motion modes reported by Pflaum et al. [36]

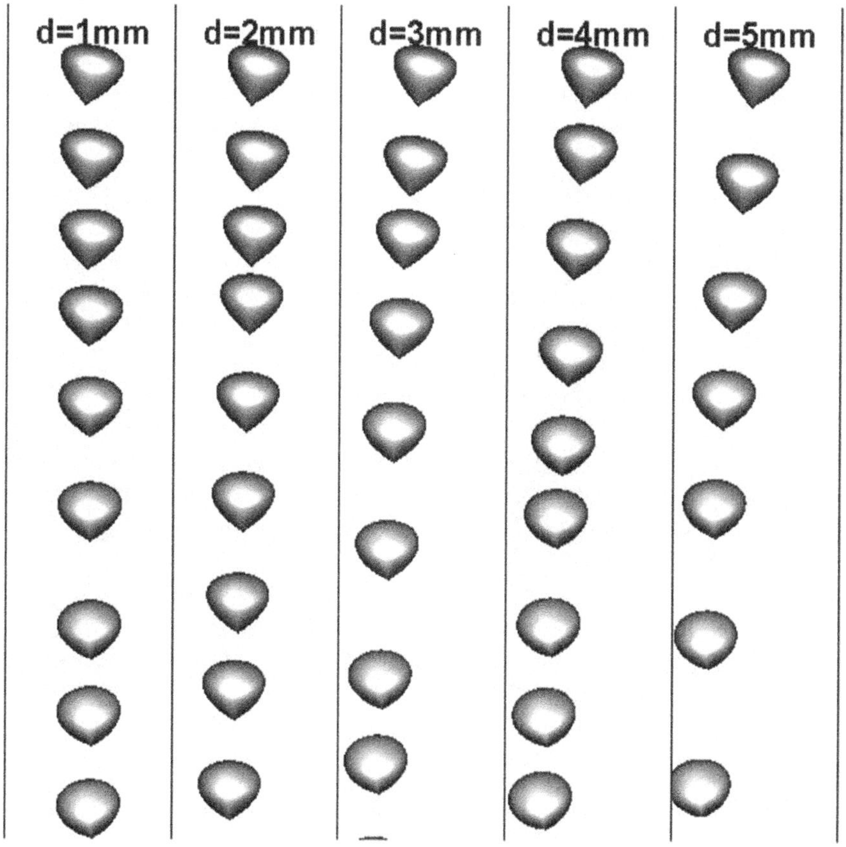

Fig. 4.51 Trajectories of the conical graupel with its initial inclination of 160° with respect to Y axis between various sizes (adapted from [8])

who performed vertical wind tunnel study of graupel hydrodynamics. These motions include rotation, pendulum swing, sailing, and tumbling. Thus our study shows that the complicated motions of highly nonspherical ice hydrometeors such as conical graupel can be successfully simulated by computational fluid dynamics techniques. All these motion modes may impact the horizontal translation of graupel.

The average horizontal speed V_H of the graupel after it reached a quasi-steady fall attitude is shown in Fig. 4.53.

In general, larger graupel tend to have larger horizontal speeds than smaller ones, although the 4 mm graupel has slightly higher V_H than the 5 mm graupel but the difference is small. Given a long time, the graupel may travel far from the origin as they fall. We see, for example, from Fig. 4.53a that V_H is about 0.31 m/s for $d = 5$ mm. Thus, in an hour the 5 mm graupel could potentially move a horizontal distance of ~1 km if it moves in the same direction away from the origin during this time period. Note this distance is translated purely due to the fall attitude of the

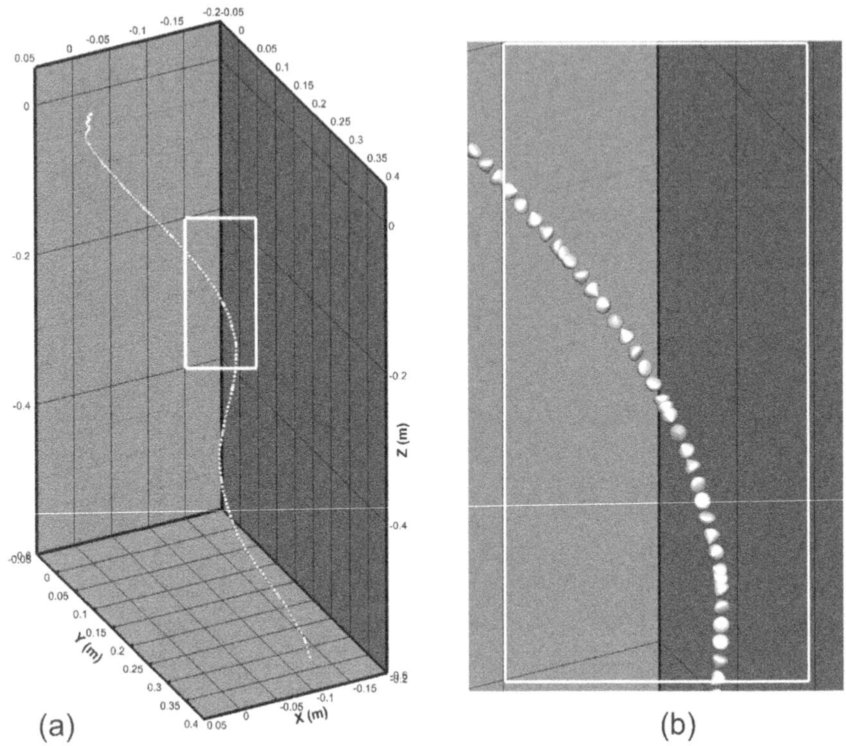

Fig. 4.52 Trajectory of the d = 5 mm graupel from t = 0 to t = 1.68 s. **a** Trajectory. **b** Enlarged view corresponding to the section indicated by the white box in (**a**) showing the fall orientations

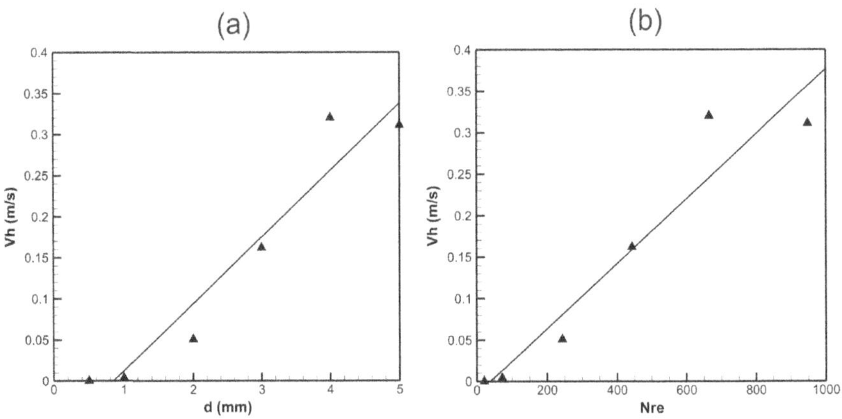

Fig. 4.53 The horizontal speed of conical graupel after they reach steady fall velocity as a function of **a** diameter, **b** Reynolds number (adapted from [9] with changes)

graupel alone and has nothing to do with the in-cloud wind speed. This implies that the distribution of ice hydrometeors such as conical graupel in clouds can be much wider than predicted by cloud microphysical models that consider only vertical fall attitudes.

Figure 4.53b shows the plot for V_H as a function of Reynolds number N_{Re}. We use linear fit for the data points in Fig. 4.53a, b as:

$$V_H = -0.06854 + 0.08126d \tag{4.28}$$

and

$$V_H = -0.01472 + 0.0003916 N_{\mathrm{Re}} \tag{4.29}$$

Although a more sophisticated 3rd-order polynomial can fit the points better, it is felt unnecessary as there is no theoretical justification of such high-order fit.

4.5.5 Drag Coefficients

The drag coefficients of the falling graupel after they reach steady state terminal fall velocity are calculated according to

$$C_D = \frac{2F_D}{\rho_a (u_\infty - V_t)^2 A_G}, \tag{4.30}$$

where F_D is the frag force acting on the graupel and A_G is the cross-sectional area of the graupel defined as a constant, $\pi(d/2)^2$. We used $(u_\infty - V_t)$ to obtain the quasi-steady Reynolds number as defined in Hashino et al. [13].

Figure 4.54 shows the comparison of the drag coefficients calculated in the present study with that of Wang and Kubicek [55], sphere, Heymsfield and Kajikawa [15] and Böhm [5]. We see that the present results are in general lower than that obtained by Wang and Kubicek [55]. This is likely due to the assumption on Wang and Kubicek [55] that the graupel fall with fixed orientation whereas the present study allows the graupel to adjust the orientation in response to the instantaneous hydrodynamic force which tends to decrease the fluid resistance. Consequently, the drag and hence the drag coefficients are generally lower in the present study.

In this regard, it is interesting to observe that the presently calculated drag coefficients are close to that of a sphere. This can be interpreted in the following manner. The reason why conical graupel drag deviates from that of a sphere is of course mainly due to its nonspherical shape. In Wang and Kubicek [55] the graupel orientation is fixed and the shape effect is also fixed, causing deviations from the sphere drag. In the present case, the graupel is allowed to adjust its orientation in response to the nonspherical shape drag and thus partially compensate the shape factor. Hence it is reasonable to expect that the drag be closer to that of a sphere.

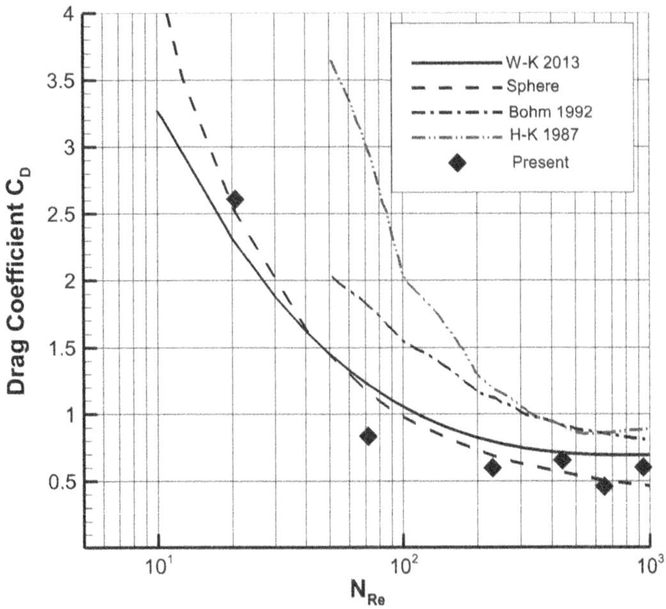

Fig. 4.54 Comparison of the drag coefficients as a function of Reynolds number calculated in the present study with some previous studies (adapted from [9] with changes)

The drag coefficients of Heymsfield and Kajikawa [15] and Böhm [5] are higher than the present results possibly because of the difference in shape parameters. The graupel in the present study are all of the standard shape and aspect ratio, and surfaces are assumed to be smooth. The data used by Heymsfield and Kajikawa [15] and Böhm [5] were from observations of natural graupel whose shapes are not uniform and surface are not smooth. Roughness on the surface normally causes higher drag if all other factors remain the same. Hence it is reasonable to expect the drag coefficient to be higher.

From Fig. 4.54, it appears to be reasonable to use spherical drag coefficient to approximate the drag coefficient of a conical particle of the same Reynolds number. However, it is to be reminded that this approximation should be used only for drag calculations. As seen in the previous section, conical particles will perform significant horizontal translations which is probably not true for a spherical particle.

In the above sections, we reported our recent numerical simulation results of the free fall of conical graupel. We demonstrated that the computational fluid dynamical models can simulate the fall of conical graupel fairly realistically. All major motion modes—rotation, pendulum swing, sailing and tumbling—are successfully and realistically simulated.

4.6 Spherical Hailstones

When graupel grow to become larger than 5 mm in dimension, they become hailstones. Hailstones are the largest hydrometeors that can fall from a precipitating cloud. Usually only thunderclouds are capable of producing such large hydrometeors [57]. Their shapes can vary from quasi-spherical to spheroidal to highly lobed and irregular. In this monograph, we will treat only two kinds: spherical hailstones and lobbed spherical hailstones.

The detailed numerical flow fields around falling hails are previously unavailable because of the difficulty in solving the complete unsteady Navier–Stokes equations for such high Reynolds number cases. With the improvement of computer performance and the advance of high performance computational fluid dynamics (CFD) software packages, it becomes practical to perform the numerical computations for such flow cases.

Here we will be treating hailstones as smooth spheres as a first approximation. Real hailstones have various degrees of roughness due to surface irregularities and they can be completely frozen or coated with liquid water. Although the flow past a sphere has been a classical subject of fluid dynamical studies (see, e.g., summary in Chap. 10, Pruppacher and Klett [41], for laminar cases and papers such as Tomboulides and Orszag [47], and Constantinescu and Squires [10], for turbulent cases), most focus on the numerical techniques. We shall focus on the results that are relevant to falling hails.

The procedure of the present computational study is different from that reported in Kubicek and Wang [27], Wang and Kubicek [55] and Chueh et al. [8, 9] who computed the unsteady flow fields around falling conical graupel. There the calculation was first performed to obtain a steady axisymmetric flow field. Then an instantaneous velocity bias is added to cause the steady field to induce the eddy shedding. Without such an instantaneous bias, the flow fields remained steady (aside from the very short transient motions) even if the transient model of Fluent was used. In the present calculation, however, we simply started the computation by using the transient model and the computed flow fields are readily unsteady. While the exact reason why this is so is unclear at the present, it is possible that the Reynolds numbers for hails falling in air are all very large (greater than several thousands) so that any reasonable numerical scheme can produce the unsteady features whereas the Reynolds numbers for falling graupel are smaller, being on the order of several hundred, which are on the borderline between steady and unsteady flow, and hence an adequate initial perturbation may be necessary to cause the numerical scheme to induce unsteady features such as eddy shedding.

4.6.1 Dimensions and Velocities of Hailstones Examined

By definition in cloud physics, a hailstone should have a diameter or the largest dimension larger than 5 mm. In the present study, we computed the flow fields for falling spherical hails of diameter 1, 2, 3, 4, 5, 7 and 10 cm. The present study assumes that the atmospheric environment is 1000 hPa and 10 °C and the hails fall at their respective terminal velocity u_∞. The terminal velocities are determined based on the empirical equation given by Knight and Heymsfield [26] for fresh hailstones in Colorado storms:

$$u_\infty = 8.445 d^{0.553} \tag{4.31}$$

where u_∞ should be in the unit of m s^{-1} and the diameter d in cm.

Table 4.6 shows the diameter, terminal fall velocity, and Reynolds number of the hailstones investigated here.

While the study reported in this section assumes the atmospheric condition at the surface level, as we intend to study the general fall behavior of hail, the results are applicable to the fall behavior of hails at higher levels because what really characterizes the flow fields is the Reynolds number. No melting is considered so that the temperature really does not matter. To use the results for hail behavior at upper levels, all we need to do is change the hail size and air density so that the same Reynolds number is reached. The flow fields of the two would look the same.

In this study, we also assume that the hail falls vertically and hence no lateral motion is considered. We had performed a few test cases to understand how significant is the hailstones lateral motion in this set of calculations and the results show that the horizontal movement of a spherical hailstone is indeed very small compared to the vertical motion, and therefore the impact of the lateral motion on the flow field is insignificant. Thus we decided to perform the calculations assuming that no lateral motion to conserve the computing resource. Since the hails are assumed to be spherical, the orientation problem associated with rotation is irrelevant. Of course the motions of real hailstones may be much more complicated (see, e.g., a short summary by Pruppacher and Klett [41], p. 441) but so far not much information is available.

Table 4.6 Dimension, Reynolds number and terminal velocity of the hails whose flow fields are calculated in this study (adapted from [7])

Diameter (cm)	Reynolds number	Terminal velocity (m/s)
1	5780	8.445
2	1.70E+04	12.39
3	3.19E+04	15.504
4	4.98E+04	18.178
5	7.04E+04	20.565
7	1.19E+05	24.771
10	2.06E+05	30.172

It will be seen that, due to the much larger Reynolds numbers of hailstones than other ice hydrometeors, the flow in the downstream (wake) region of a falling hailstone is completely turbulent. The turbulent wake characteristics will be the main focus of our discussions below.

4.6.2 Flow Characteristics

In the following paragraphs, we will use the cases of a small hail ($d = 1$ cm) and a large hail ($d = 5$ cm) to describe the flow field characteristics of falling spherical hailstones. Then we will summarize and inter-compare all cases. We will focus on the descriptions of main characteristics including as flow pattern, pressure distribution, z-velocity, and vorticity that are relevant to hail growth.

4.6.2.1 $d = 1$ cm Case (Pea-Size Small Hail)

The left panel of Fig. 4.55 shows a randomly selected frame of the computed streamtrace field around a falling spherical hail of 1 cm diameter ($N_{\mathrm{Re}} = 5780$). As time goes on, the recirculation bubble sways slightly but its size remains quasi-constant. The eddy shedding continues downstream beyond the bubble, forming a wavy vortex street. Outside of this region, the flow is mainly laminar. The right-panel figure shows another snapshot at a different time. It is clear that the recirculation bubble size is about the same although the eddy circulation inside it has changed substantially. The flow in the upstream half is basically steady all the time. Such flow characteristics are consistent with similar calculations by others (e.g., [42]) for flow past a sphere at Reynolds number 5000.

Figure 4.56 shows the velocity vectors and the pressure distribution on a central xy-plane corresponding to that in the left panel of Fig. 4.55. Remember that the vectors are the 2-D projection of the 3-D vectors on this plan and the crowdedness of vectors in the center of the figure is due to the high density of grids there. As it is shown, the recirculation bubble is a low pressure and generally low speed (except in the far downstream eddies) region. The high speed eddies near the downstream end of the bubble are associated with the lowest pressures. On the other hand, there are a few high pressure pockets immediately downstream of the high speed eddies associated with the lowest pressures. As usual, the upstream side of the falling hail is characterized by a high pressure region.

Figure 4.57 shows the distribution of the z-velocity. Maximum positive velocity zone is located at slightly upstream of the equator and extends toward downstream. In 3-D, this zone would take the shape of a Champaign flute glass. Inside it is the recirculation bubble which, as mentioned previously, is a generally low speed region but contains a few high downward speed pockets at downstream end. The upstream region near the falling hail is a low velocity region consistent with the name "front stagnation".

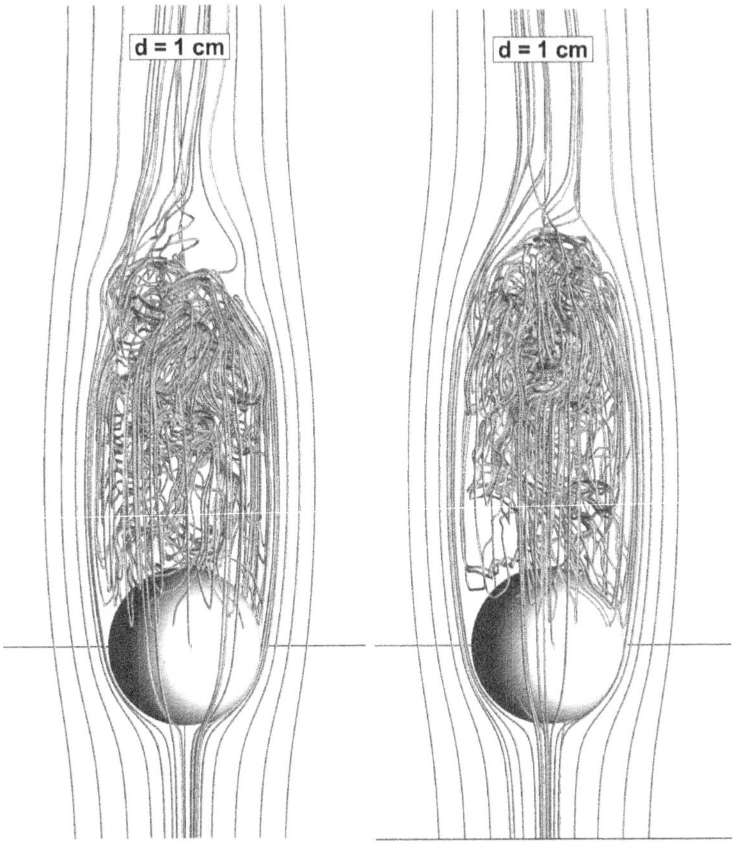

Fig. 4.55 Streamtrace patterns around a smooth spherical hailstone of 1 cm diameter falling at $N_{\mathrm{Re}} = 5780$

The huge gradient of vertical velocity near the equator implies that the vorticity will be very large here. This is indeed the case as shown in Fig. 4.58.

Here we see that the main maximum vorticity region is also located in the region slightly upstream of the "equator" of the hail, obviously due to the very large shear. There are a few high vorticity pockets in the downstream end of the recirculation bubble also. Inside the recirculation bubble the vorticity is generally low. The upstream has very small vorticity and the flow is thus basically irrotational except very near the hail surface. Note that we show here only the vorticity magnitude and not its vector properties.

While the flow characteristics described above are all obviously unsteady, it is interesting to note that the "unsteadiness" of these characteristics is quasi-steady, i.e., their patterns and magnitudes do not change very much with time and all occur at the same general locations.

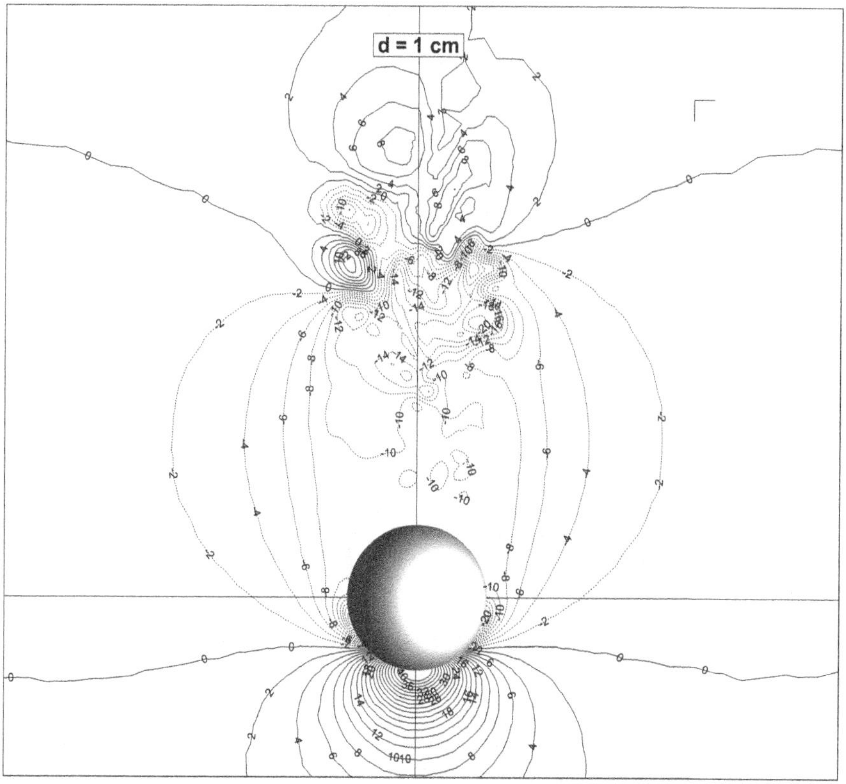

Fig. 4.56 The pressure distribution around a smooth spherical hailstone of 1 cm diameter falling at $N_{Re} = 5780$. Dashed curves represent negative pressure deviation

4.6.2.2 *d* = 5 cm Case (Golf Ball Size Large Hail)

According to the definition of the National Weather Service of US National Oceanographic and Atmospheric Administration (NOAA), a storm is called severe if it produces hails with size greater than 1 inch (about 2.54 cm) in diameter (see https://www.spc.noaa.gov/faq/#4.2). Thus most of the hail cases we examine in this study belong to those of severe storms. A hailstone of 5 cm diameter ($N_{Re} = 70{,}400$) is about 2 inches which is larger than a standard golf ball. Hailstones of this size, if falls on the roof of a house or an automobile, may produce substantial damage.

Figure 4.59 shows a randomly selected snapshot of the streamtrace pattern. We see that the size of the recirculation bubble is much smaller than that in Fig. 4.55. This is consistent with experimental results of similar Reynolds number range that are included in Van Dyke [48] that the larger RE in this range cause the bubble to become smaller. While the recirculation bubble becomes smaller, the whole wake region becomes more turbulent as it should be with the increasing Reynolds number.

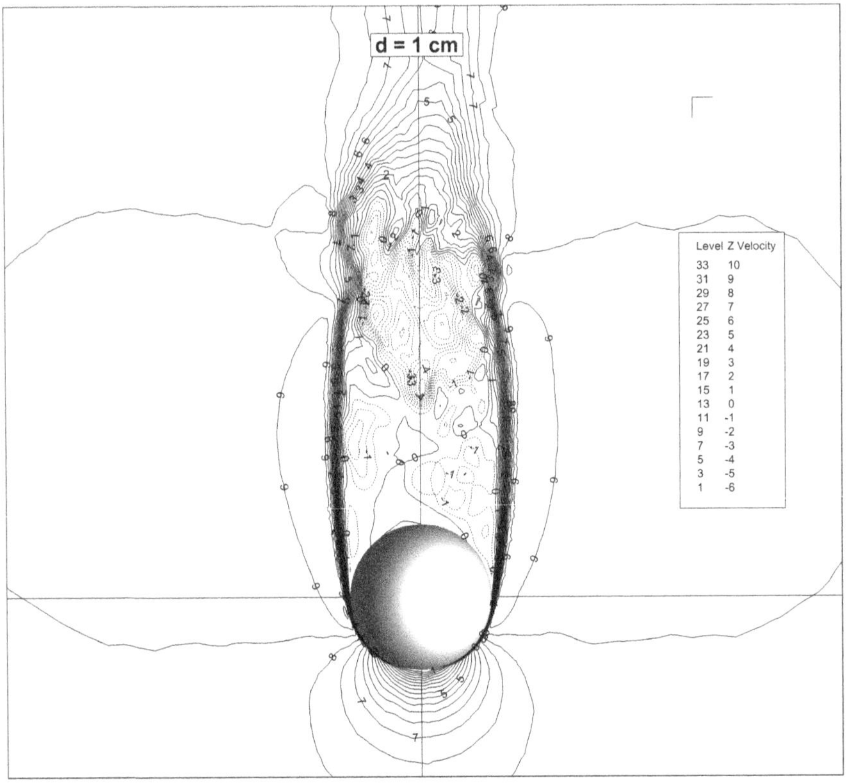

Fig. 4.57 The z-velocity distribution around a smooth spherical hailstone of 1 cm diameter falling at $N_{\mathrm{Re}} = 5780$

The pressure distribution corresponding to that in Fig. 4.59 is shown in Fig. 4.60. Comparing this with the $d = 1$ cm case in Fig. 4.56, we note that the lowest pressure regions in the recirculation bubble are now much closer to the hail surface as the Reynolds number increases. The difference between the maximum and minimum pressure is now much greater than that in Fig. 4.56 (but note that the pressure scales are different in these two figures).

Figure 4.61 shows the corresponding vertical velocity distribution. One notable difference from that of the 1 cm hail (Fig. 4.57) is that the high velocity region along the flank of the recirculation bubble is no longer laminar but broken up into several high velocity centers, obviously due to the more turbulent separation of the boundary layer. The maximum "negative" velocity center in the bubble is now located much closer to the hail surface, which is possibly favorable for small cloud droplets or aerosol particles trapped in the wake to descend to and collide with the hail (see Chap. 6 of [57] for wake-trapping of aerosol particles). But whether this would really happen is currently uncertain and will be studied further.

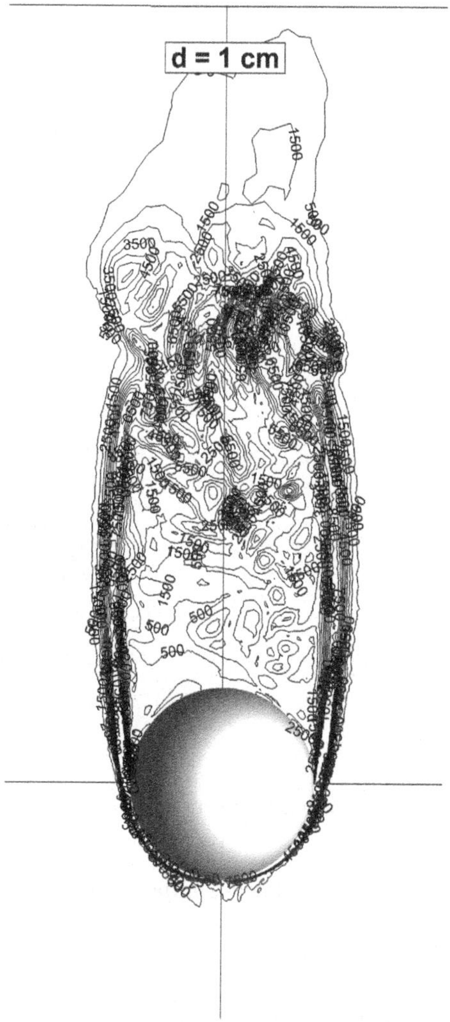

Fig. 4.58 The vorticity magnitude distribution around a smooth spherical hailstone of 1 cm diameter falling at $N_{\mathrm{Re}} = 5780$

Figure 4.62 shows the corresponding vorticity magnitude distribution. Unlike the long stretched high vorticity flute flanking the recirculation bubble, the high vorticity is confined to just the vicinity of the hail surface. Note, however, the maximum vorticity magnitude in the present case is actually **5 orders of magnitude** greater than that of the 1 cm case in Fig. 4.58. In contrast, the recirculation bubble is now much more turbulent than that in Fig. 4.58, as much higher vorticity fluid has been transported into the bubble.

Flow properties as described above will have important impacts on the hail cloud physical behavior such as ventilation effect and collision with small particles.

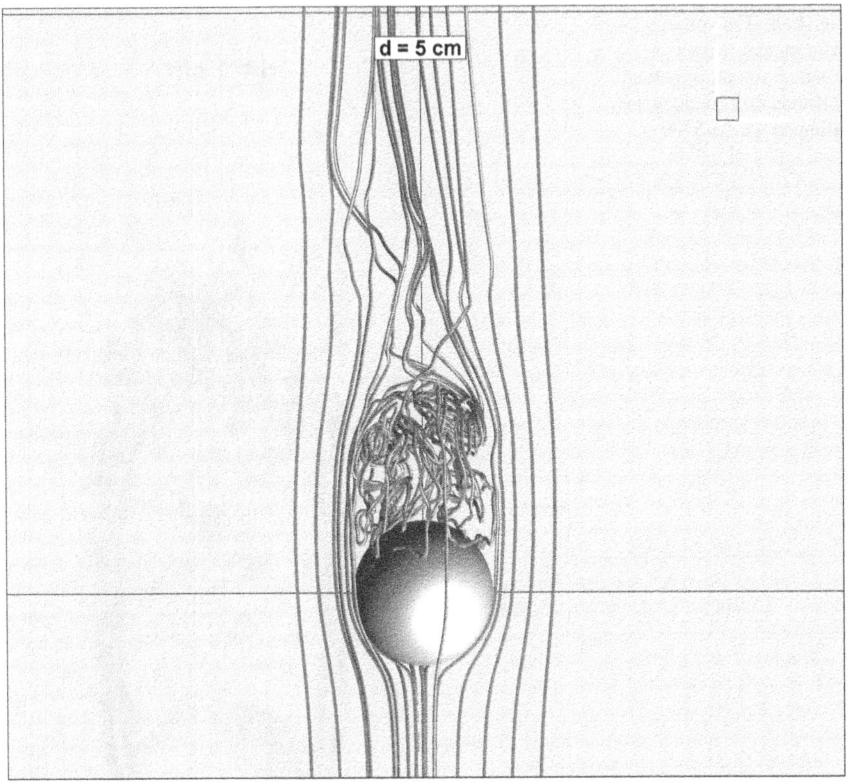

Fig. 4.59 Streamtrace pattern around a spherical hailstone of 5 cm diameter falling at $N_{\mathrm{Re}} = 70{,}400$

4.6.2.3 Flow Filed Comparison

The above paragraphs gave two examples among the cases studied. Figures 4.63 and 4.64 show the summary of the flow characteristics for the 5 cases of $d = 1, 3, 5, 7$ and 10 cm. For brevity, we only show the streamtrace patterns and the vertical velocity distribution. Figure 4.63 shows the comparison of the streamtrace pattern among the 5 cases. Clearly, the recirculation bubble size **decreases** systematically with Reynolds number. Also, the wake shape changes from quasi-spheroidal to pyramidal as the Reynolds number increases. The motion in the wake region becomes more turbulent although it cannot be easily shown by these figures. The turbulent nature is better shown by the z-velocity distribution given below.

Figure 4.64 compares the z-velocity fields for all 7 cases. The most obvious feature is that the minimum (i.e., maximum "negative") velocity center in blue color is located closer and closer to the rear of the hail surface as the hail size increases. The increasingly turbulent wake flow can be clearly seen. Another notable feature is the transition of the maximum z-velocity zone with hail size. At $d = 1$ cm, the maximum velocity zone (red) flanking the recirculation bubble is quasi-steady and laminar and,

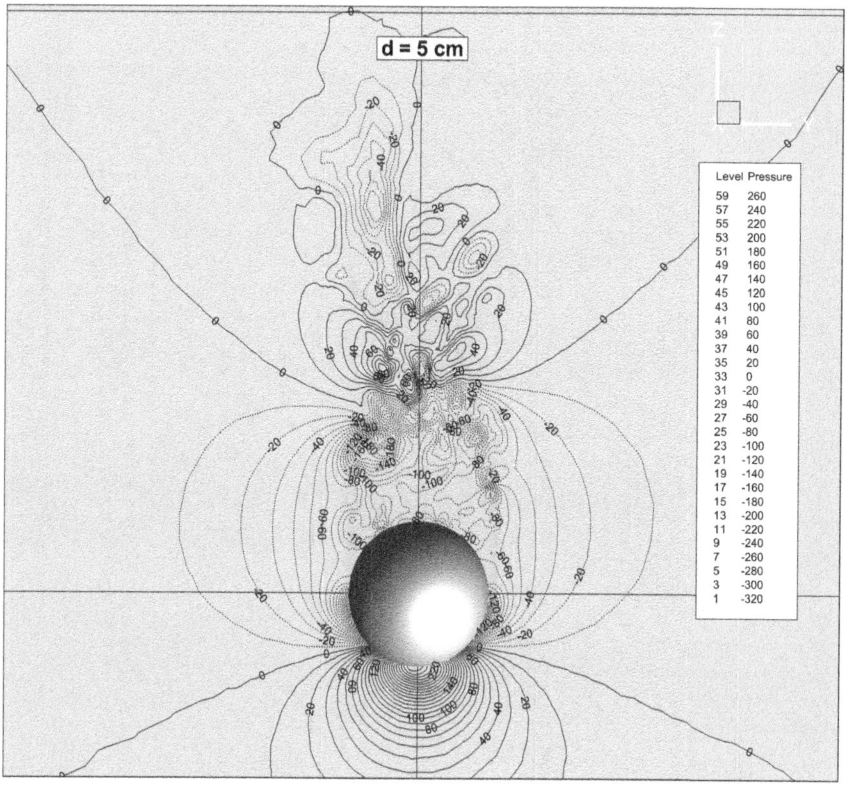

Fig. 4.60 Pressure pattern around a spherical hailstone of 5 cm diameter falling at $N_{\mathrm{Re}} = 70{,}400$

as mentioned earlier, it has a shape of a Champaign flute. Within the flute there is a single velocity maximum located near the equator. Beginning at $d = 2$ cm, however, this maximum zone not only becomes smaller in size and behaves less steady, the single maximum structure also breaks up into multiple maximum centers. In 3-D, these centers will resemble a net surrounding the recirculation bubble. As the hail size increases, the net breaks into rings that will be shed similar to the eddy shedding phenomenon. Figure 4.65 shows an example of this **ring-shedding** phenomenon in 3-D (right panel) and its corresponding 2-D plot (left panel) for $d = 5$ cm. It is seen that the rings may tilt at certain angle and interfere with others.

Figure 4.66 shows the variation of the eddy length to hail diameter ratio (averaged over time, as the length fluctuates slightly) as a function of the hail diameter. The eddy length is defined as the vertical distance from the rear stagnation point of the hail surface to the point beyond which no negative z-velocity occurs. The ratio varies smoothly. As the hail diameter increases, the ratio decreases. The data can be fitted by the following empirical formula:

$$(L/d) = 2.21 d^{-0.621} \tag{4.32}$$

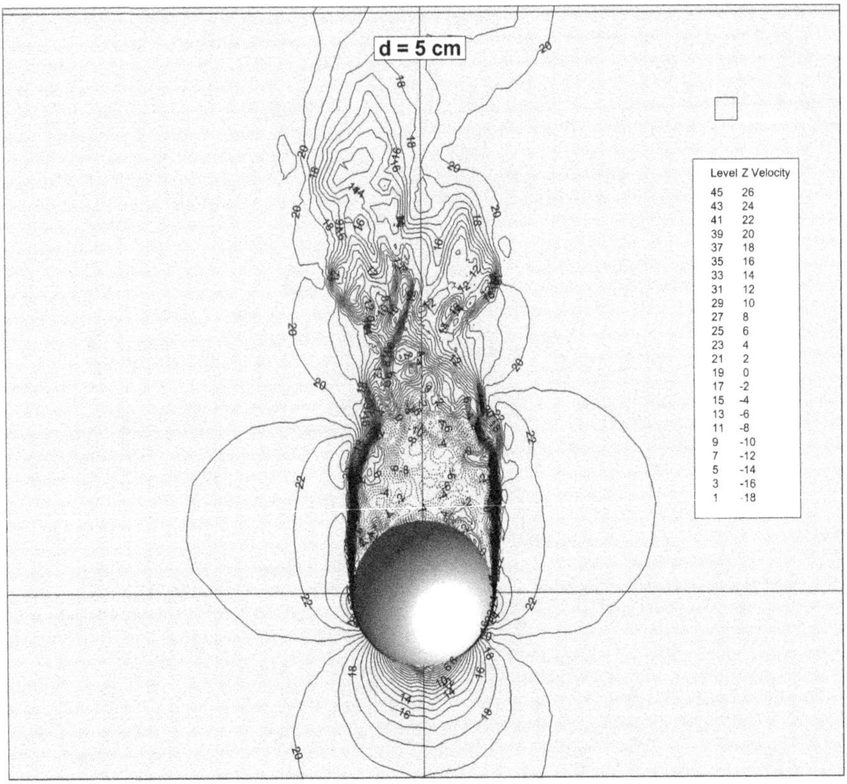

Fig. 4.61 Z-velocity pattern around a spherical hailstone of 5 cm diameter falling at $N_{\mathrm{Re}} = 70{,}400$

Figure 4.67 shows the same ratio versus the hail Reynolds number. The data can be fitted by the following empirical formula:

$$(L/d) = 70.75(N_{\mathrm{Re}})^{-0.4} \tag{4.33}$$

4.6.3 Drag Coefficients

Figure 4.68 shows the computed drag coefficient as a function of hail diameter. The data points can be fitted by the following formula very closely:

$$C_D = 0.2984 + 0.1177d - 3.7985 \times 10^{-2}d^2 + 3.8594 \times 10^{-3}d^3 - 1.3305 \times 10^{-4}d^4 \tag{4.34}$$

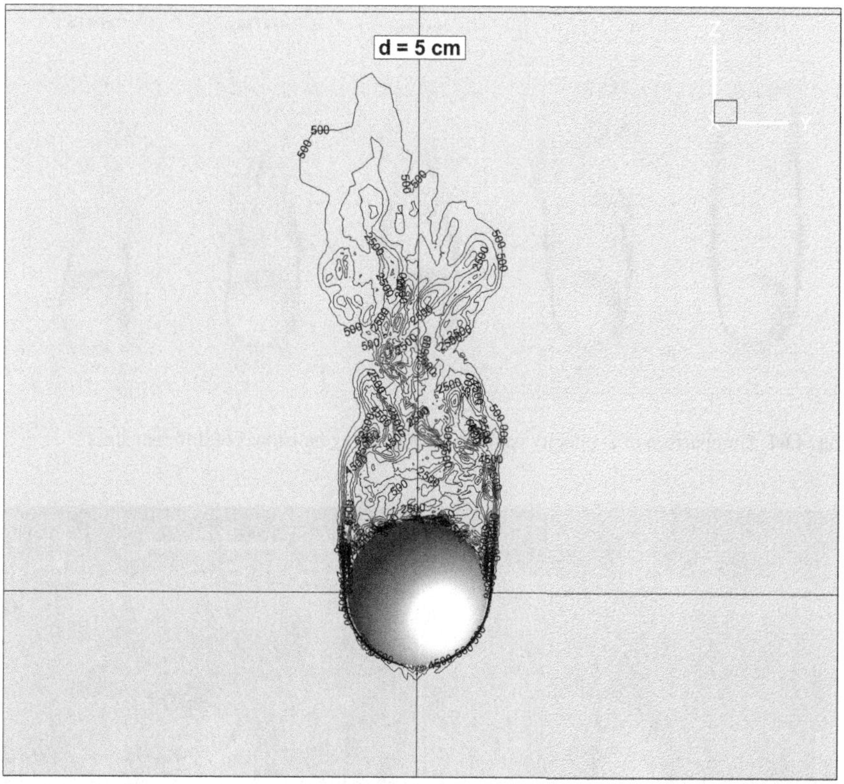

Fig. 4.62 Vorticity magnitude pattern around a spherical hailstone of 5 cm diameter falling at N_{Re} = 70,400

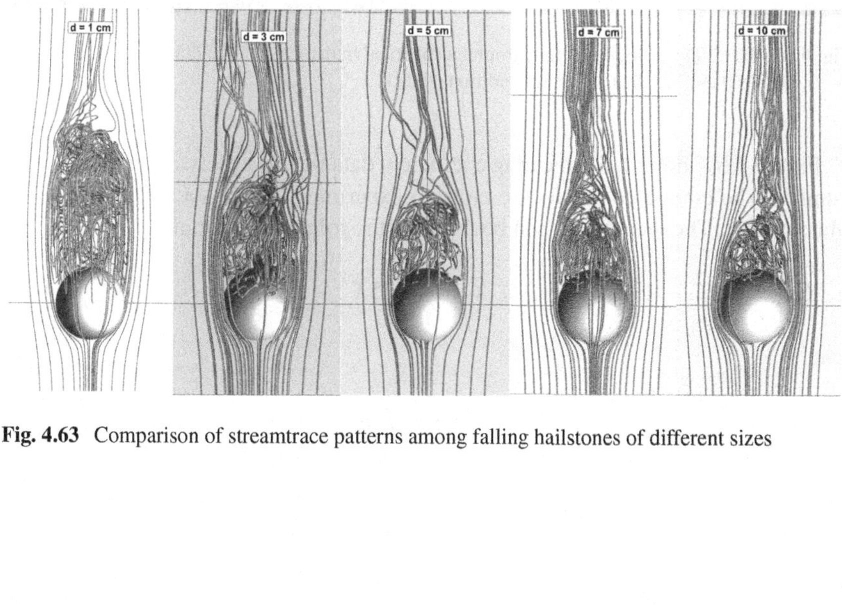

Fig. 4.63 Comparison of streamtrace patterns among falling hailstones of different sizes

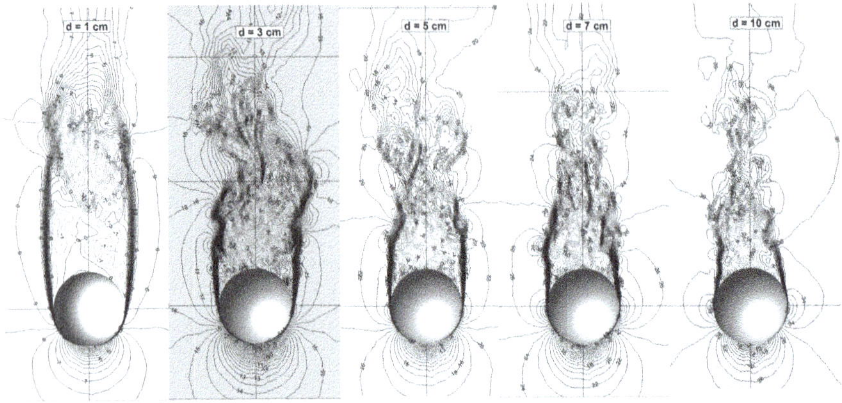

Fig. 4.64 Comparison of z-velocity patterns among falling hailstones of different sizes

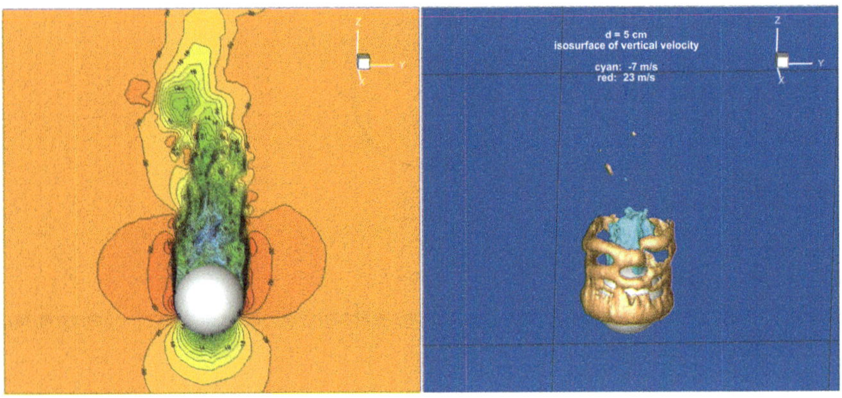

Fig. 4.65 The 2-D z-velocity pattern around a spherical hailstone of 5 cm diameter falling at $N_{\mathrm{Re}} = 70{,}400$ (left) and it's 3-D structure rendition (right)

Figure 4.69 shows the same drag coefficient data but the abscissa is hail Reynolds number. The computed results are close to experimental data such as presented in Massey [30]. The data points can be fitted by the following formula:

$$C_D = 61.62 - 49.44X + 13.07X^2 - 0.835X^3 - 0.141X^4 + 0.016X^5 \quad (4.35)$$

where

$$X = \log_{10} N_{\mathrm{Re}} \quad (4.36)$$

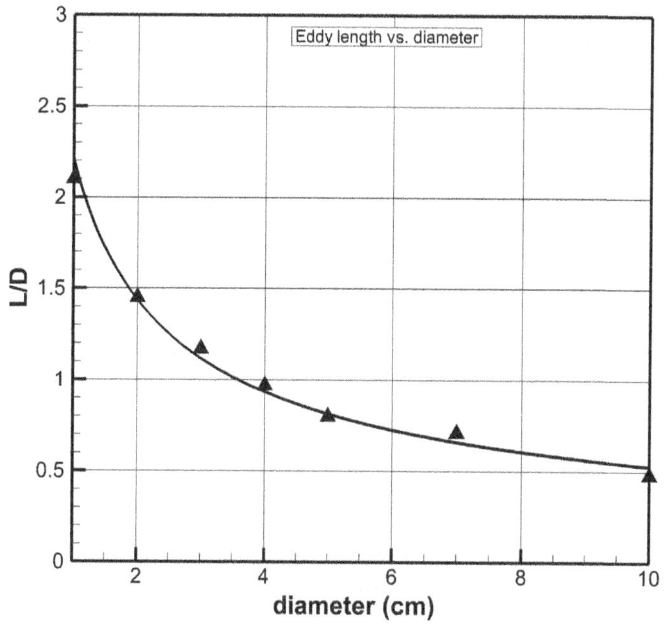

Fig. 4.66 The eddy length in the wake as a function of spherical hailstone diameter (adapted from [6] with changes)

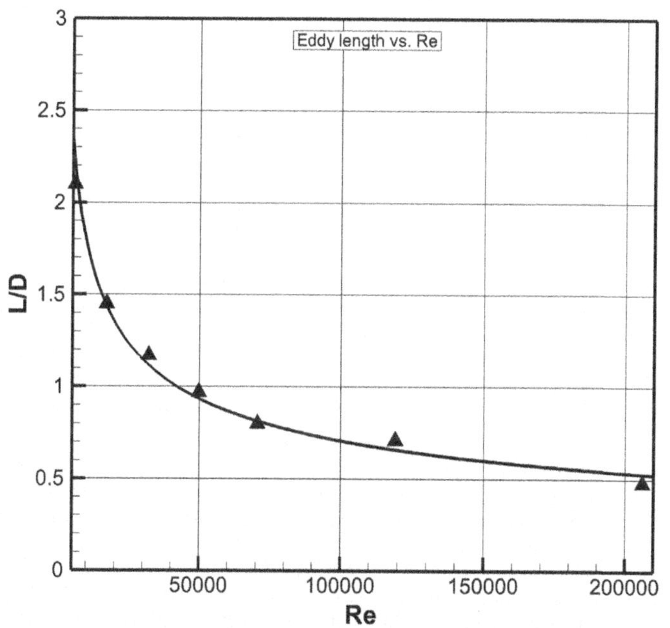

Fig. 4.67 The eddy length in the wake as a function of spherical hailstone Reynolds number (adapted from [6])

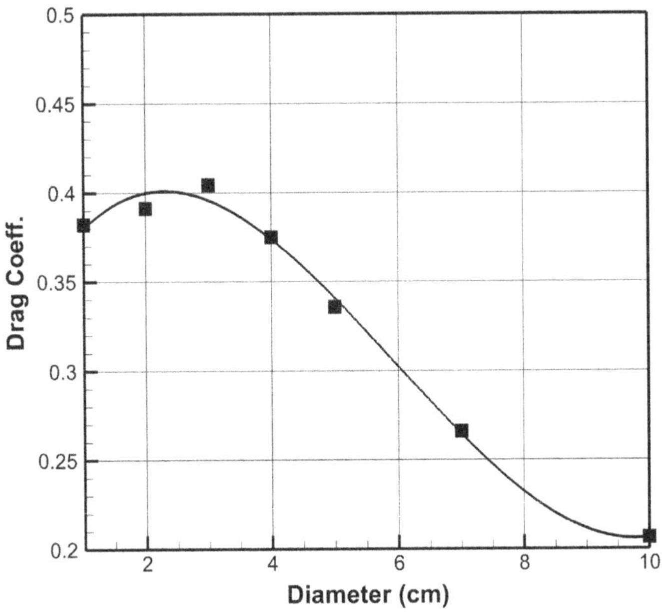

Fig. 4.68 Drag coefficient versus diameter for spherical hailstones (adapted from [6])

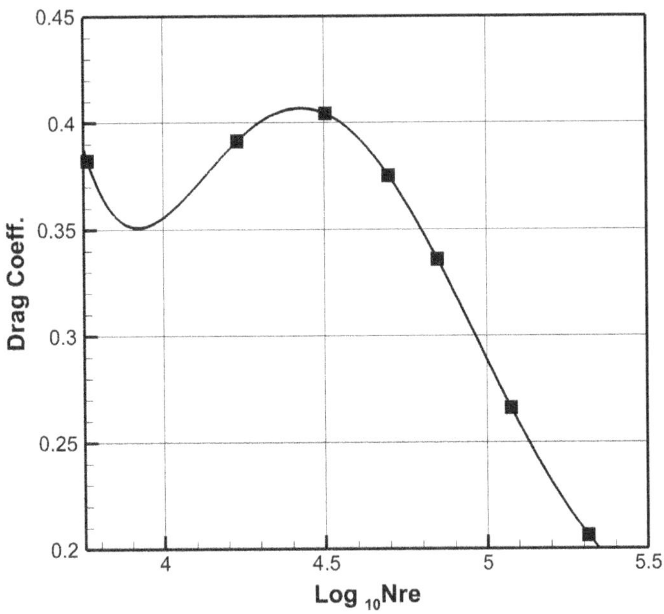

Fig. 4.69 Drag coefficient versus Reynolds number for spherical hailstones (adapted from [6])

4.7 Lobbed Hailstones

The hailstones we discussed in the previous section are spherical hailstones with smooth surface. As we mentioned in Chap. 1, many hailstones are lobbed. Some of them have lobbed distributed more or less uniformly over the surface while others (especially the very large ones) have lobes distributed asymmetrically. An example of hailstones including some lobbed ones is shown in Fig. 1.11. The lobbed stones in this sample are the quasi-uniform distribution kind. In the following discussion, we will deal with this kind of lobbed stones. The materials of the following discussion are taken from Wang et al. [53].

4.7.1 Mathematical Formulation for Lobbed Hailstones

To describe the shape of these lobbed hailstones, we use a set of mathematical equations based on SMOSS [51, 52] to describe the shapes of lobed hailstones used in this calculation. The equations in Cartesian coordinates collectively giving the 3D surface points of the stones are shown in the following:

$$x = \cos(w)\cos(l)\left\{p\left(1 - \cos^2(qw)\right)^r\left(1 - \cos^2(ql)\right)^r + s\right\}, \tag{4.37}$$

$$y = \sin(w)\cos(l)\left\{p\left(1 - \cos^2(qw)\right)^r\left(1 - \cos^2(ql)\right)^r + s\right\}, \tag{4.38}$$

$$z = \sin(l)\left\{p\left(1 - \cos^2(qw)\right)^r\left(1 - \cos^2(ql)\right)^r + s\right\}, \tag{4.39}$$

where the variable w varies from 0 to 2π and l from -0.5π to 0.5π, respectively. The other parameters (i.e., p, q, r and s) that we use in the set of mathematical formulation are adjustable input parameters to simulate the shape and size of the lobed hailstones. By changing values of p, r and s we can create hailstones with long or short lobes. Figure 4.74 shows the lobed hailstones created by this method (the first row shows the short lobed while the second row long lobed).

The values of the parameters for these shapes are given in Table 4.7.

Table 4.7 The parameters used in the mathematical expressions (4.37)–(4.39) (adapted from [53] with changes)

Parameter in equation	Long lobed hailstones	Short lobed hailstones
P	0.5	1
R	1	2
S	2	8

The last parameter q contained inside the cosine square terms serves to control the number of lobes to represent various degrees of roughness due to surface irregularities. For example, for both long and short lobed cases of $q = 6$ (see the right most column panels labeled as 6uv in Fig. 4.70), there are 6 latitudinal circles from North to South Pole of the hailstone, each of which has 12 lobes, and hence the hailstone has 72 lobes in total. For the cases labeled as 4uv as shown in the middle column panels in Fig. 4.70, the parameter q now changes to 4, resulting in the appearance of 4 latitudinal circles, each of which has 8 lobes and thus the hailstone has 32 lobes in total. Likewise, for 2uv cases where $q = 2$ (see the left most panels in Fig. 4.70), the hailstone has 8 lobes in total. It is important to note that the "uv" for all in the labelled names of the lobed hailstones has no particular meaning, but is just simply a symbolic label to make the labelled abbreviations look long enough to be identified quickly.

Note that the lobes generated in this manner are not of uniform size. The lobes near the "equator" of the hailstone are the largest and the size decreases with "latitude". Thus those lobes near the pole tend to be smaller. However, the lobes in natural hailstones are not uniform either, so although it is possible to make the lobes more uniform, we feel that the present method still gives a reasonable approximation of the hailstone roughness to the first degree.

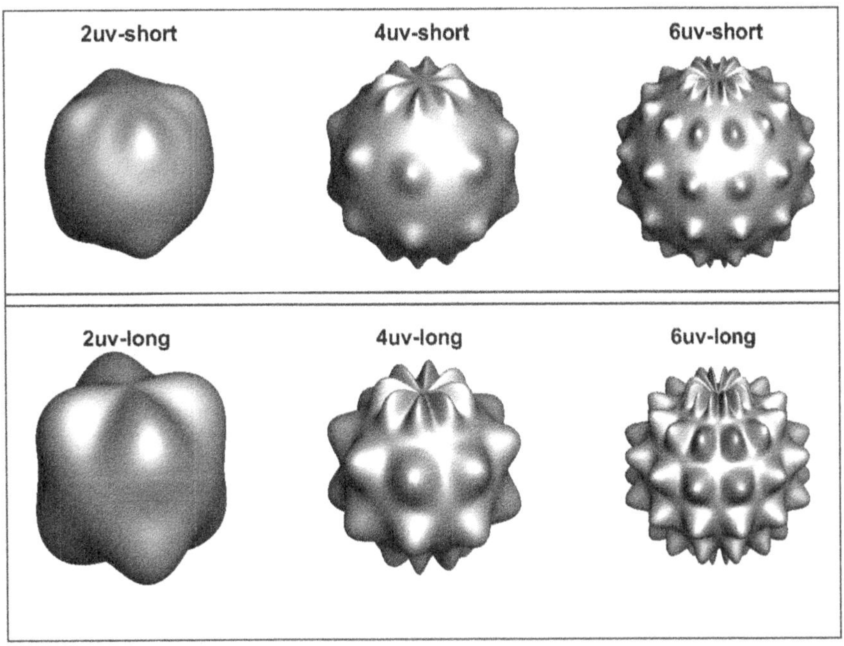

Fig. 4.70 Six examples of simulated hailstones generated by Eqs. (4.37)–(4.39). The numerical values of the parameters in these equations are given in Table 4.7 and the text (adapted from [53] with changes)

In order to achieve physically representative flow fields around the falling lobed hailstones, the Large Eddy Simulations (LES) option within Fluent is selected for turbulence modeling in the present study. Unlike Reynolds Averaged Navier–Stokes Simulations (RANS) method in which larger eddies are not resolved because of time average, LES allows us to capture more minute eddies. The simulation using K-omega (which belongs to RANS category) turbulence modelling produces relatively small turbulence (or fewer eddies) and therefore appears more laminar in general. On the other hand, the LES result shown in the right panel appears more turbulent due to the presence of many smaller eddies. For more details of the two kinds of turbulence modeling, readers are referred to Pope [40]. All of the simulations in Cheng and Wang [6] and presented in the previous section used LES option for their modeling although not mentioned in their paper.

The present study assumes that the atmospheric environment is 900 hPa and 10 °C. This is relevant to the situation when hailstones fall below the cloud base. The hailstones we are considering here are fairly large and they can fall for 2–3 km without substantial melting as the ventilation-enhanced cooling would prevent quick melting. The flow field is mainly a function of Reynolds number and hence the flow characteristics reported here will be representative for lobed hailstone in general because of the similar Reynolds number range. We also assume that during the simulations there is no change in the shapes of lobed hailstones while they plummet towards the ground.

In the case of spherical hailstones, we used Eq. (4.39) to determine their terminal velocities. But that formula was established assuming the hailstone density is about 0.44 g cm^{-3}, which is for fresh low density hailstones produced in Colorado as reported by Knight and Heymsfield [26]. This density is lower than many hailstones that have density closer to 0.9 g cm^{-3} and it doesn't consider the impact of air density and drag coefficient on the terminal velocities. In this lobed hail study, we choose another formula proposed by Pruppacher and Klett [41] for a roughly spherical hailstone falling at terminal velocity:

$$u_\infty = 0.36\left(\frac{\rho_H d}{C_D \rho_a}\right)^{0.5}, \tag{4.40}$$

where ρ_H is the density of the hailstone and C_D is the drag coefficient. It is a function of air density, hailstone density, hailstone diameter as well as drag coefficient. And air density for this formula is 1.2922×10^{-3} g cm^{-3} and hailstone density is 0.9167 g cm^{-3}.

4.7.2 Characteristics of the Flow Fields Around Falling Lobed Hailstones

In order to compare our lobed results with those of the spherical hails presented in the last section, we computed the flow fields for falling lobed hailstones of the same diameters, namely 1, 2, 3, 4, 5, 7, 10 cm. The diameters of the lobed hailstones investigated herein include the lobes. This implies that the lobed spheres actually have less mass than the corresponding smooth spheres. Consequently, one might think that, when their Reynolds numbers are the same, a lobed hailstone actually falls vertically at a higher terminal velocity than the corresponding smooth one. On the other hand, Table 4.8 shows that most of the lobed hailstones considered here are close to the spherical one in mass/volume ratio and therefore, for simplicity, we assume that they have the same corresponding terminal velocities.

The only two cases that have mass/volume ratios substantially different from 1.0 are 2uv-long and 6uv-long cases. The former doesn't look like a spherical one but looks more like a cubical molar tooth. Its resulting flow fields, as will be shown below, are therefore going in a way different than those of the other sphere-like lobed hailstones. The latter case (6uv-long) has the relatively low ratio of 0.77, because it has more holes or cavities on the surface, leading to the small volume eventually. Here, even though both of the cases have the ratio substantially deviating from 1, for the purpose of simplicity, we set that the terminal velocities of both cases are the same as the spherical one with equal diameter.

Table 4.9 shows the diameter, terminal fall velocity, and Reynolds number of the hailstones investigated here.

In the subsequent subsections, we will first discuss the flow characteristics of short lobed hailstones and then those of long lobed ones. We will focus not on the

Table 4.8 Mass/Volume ratio of (V_{LO}) hailstone to the spherical one (V_{SP}) of the equal diameter where V_{SP} is the spherical volume equal to $4/3\pi r^3$ (adapted from [53] with changes)

Vol. ratio	2uv-short	2uv-long	4uv-short	4uv-long	6uv-short	6uv-long
V_{LO}/V_{SP}	0.999324	1.229397	0.996796	0.93479	0.904591	0.770601

Table 4.9 Dimension, Reynolds number and terminal velocity of the hailstones whose flow fields are calculated in this study (adapted from [53] with changes)

Diameter (cm)	New terminal velocity (m/s)	New Reynolds number
1	13.97	9.56E+03
2	20.65	2.83E+04
3	25.94	5.33E+04
4	30.76	8.42E+04
5	35.80	1.22E+05
7	44.74	2.14E+05
10	56.65	3.88E+05

size of different hailstones but on the effects of different lobe lengths (long and short lobes) and different numbers of lobes (2uv, 4uv, and 6uv cases) that are relevant to hailstone growth. Other flow properties such as the details of instability and vortex shedding will be analyzed and published in the future. In both long and short lobed cases, we shall use hailstones with diameters 1 and 5 cm as examples to illustrate the flow characteristics.

4.7.2.1 Short Lobed Cases

Figure 4.71 shows randomly selected frames of the computed streamtrace fields around the three falling short lobed hailstones of 1 cm diameter falling at $N_{Re} = 9560$. It is seen that all three cases have cocoon-shaped recirculation bubbles in the wake. They are almost equal in vertical length, indicating that the number of lobes has little influence on this feature. On the other hand, the streamtrace field in the 6uv case shows more turbulent features than that in 4uv case, which in turn looks more turbulent than that in 2uv case, indicating that **more lobes do cause more turbulence** in these cases. Figure 4.72 shows the vertical velocity fields (positive upward and negative downward).

In all three cases, the maximum positive velocity zone is always located slightly upstream of the equator and extends towards downstream, which in 3 dimensions would take the shape of a champagne flute glass as described in Cheng and Wang [6]. The size of the champagne flute in the 6uv case is greater than that in the 4uv case, which is in turn greater than that in the 2uv case. Thus the size of maximum velocity zone increases with the number of lobes. On the other hand, the region of high downward speed inside the recirculation bubble shrinks gradually as the number of lobes increases. Figure 4.73 shows the three pressure distributions on a central xy-plane corresponding to Figs. 4.71 and 4.72.

Figure 4.74 shows the three fields of vorticity magnitude. As the number of lobes increases, there is more chance of getting some isolated, small high vorticity magnitude spots behind a hailstone with more lobes.

Next we discuss the case of the 5 cm diameter hailstone. Figure 4.75 shows randomly selected snapshots of streamtrace pattern for three short lobed hailstones of d = 5 cm.

There are two notable differences between this and Fig. 4.71. First of all, the recirculation bubble is now smaller than the corresponding hailstone in Fig. 4.71. This indicates that, as the Reynolds number increases, the recirculation bubble size decreases in general which is the same as that we discussed in the spherical hailstone case. But there is an additional difference—as the number of lobes increases, the recirculation bubble becomes smaller, which is quite unlike that in Fig. 4.71 where the bubble sizes are roughly the same. This is consistent with some observation that, at a suitable Reynolds number range, surface roughness can effectively reduce the formation of the recirculation in certain Reynolds number ranges [1].

Figure 4.76 shows the three distributions of the vertical velocity.

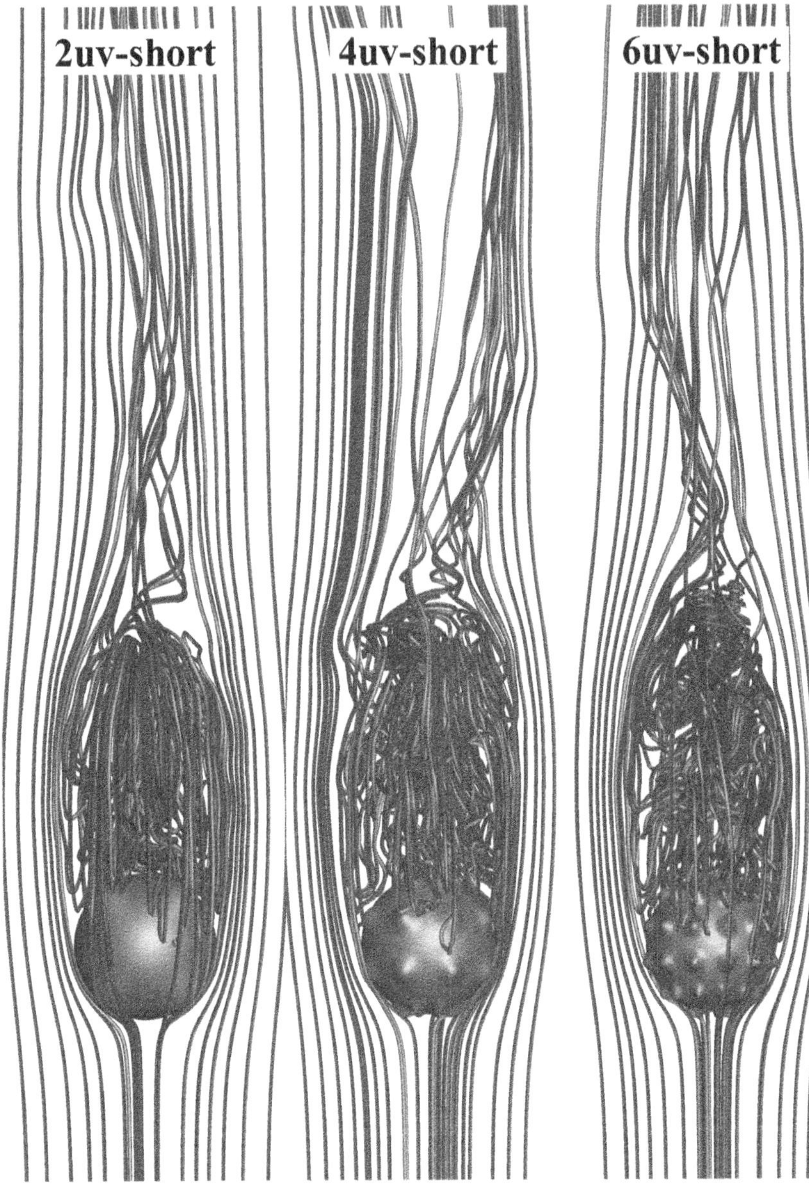

Fig. 4.71 A randomly selected snapshot of the streamtrace pattern of the computed flow field around three short lobed falling hailstones of 1 cm diameter ($N_{Re} = 9560$) (adapted from [53] with changes)

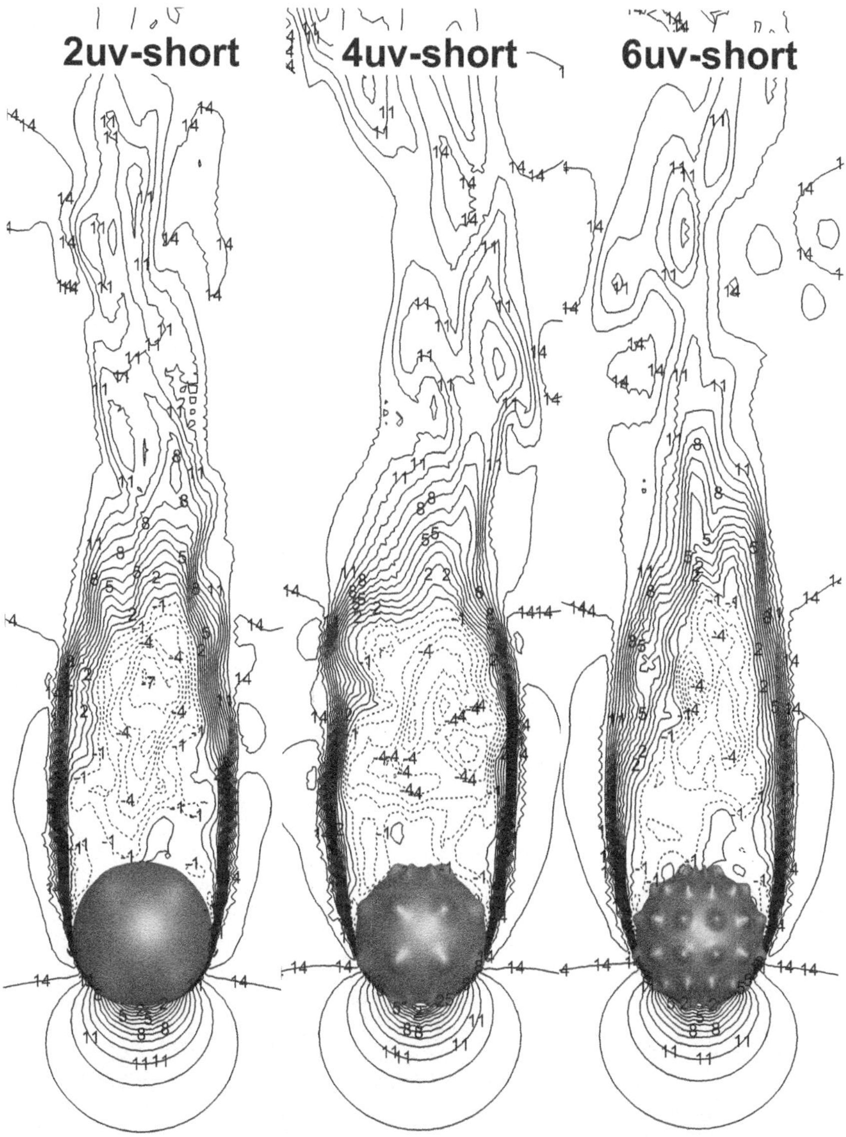

Fig. 4.72 The z-velocity field in the central yz-plane corresponding to Fig. 4.71 (adapted from [53] with changes)

Although the regions near the front surfaces of the three hailstones look similar to each other, the regions of maximum negative velocity are located much closer to the rear surface of a hailstone as the hailstone has more protrusions on its surface. This is consistent with the feature that the more lobes on a hailstone, the smaller the recirculation bubbles. Figure 4.77 shows the pressure distributions corresponding to

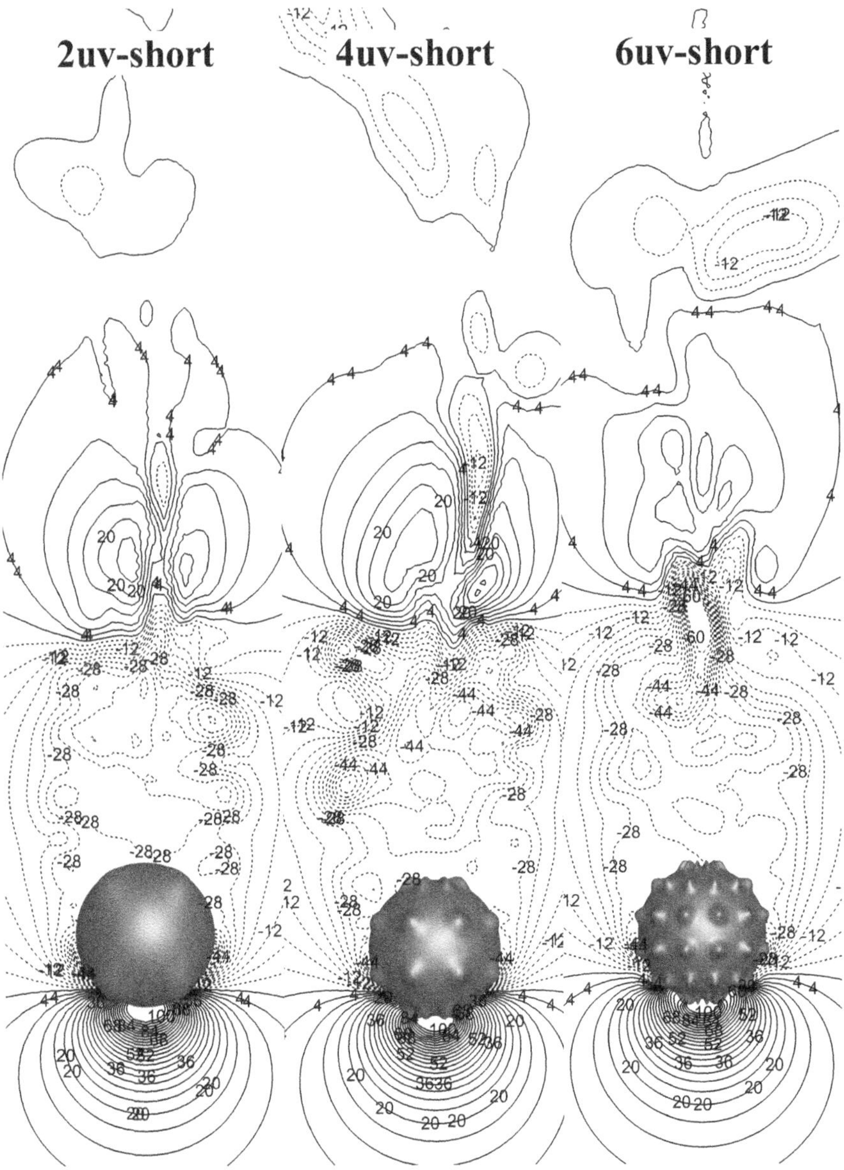

Fig. 4.73 The pressure field in the central yz-plane corresponding to Fig. 4.71 (adapted from [9]). This figure indicates that, as the number of lobes rises, the region of low pressure (behind the hailstone) becomes larger (adapted from [53] with changes)

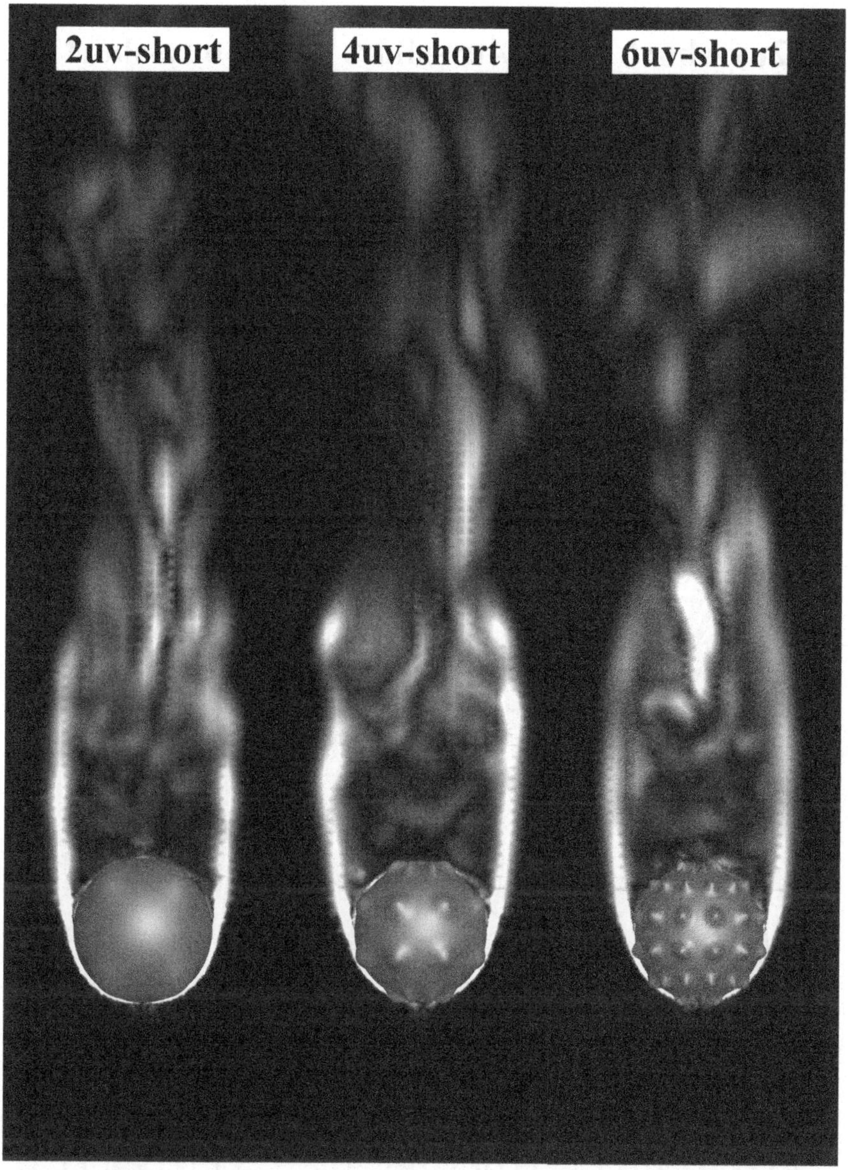

Fig. 4.74 The Vorticity Field in the central YZ-plane corresponding to Fig. 4.71. The maximum (red) and minimum (blue) vorticity magnitudes shown in these contours are 500 and 50,000 (1/s) (adapted from [53] with changes)

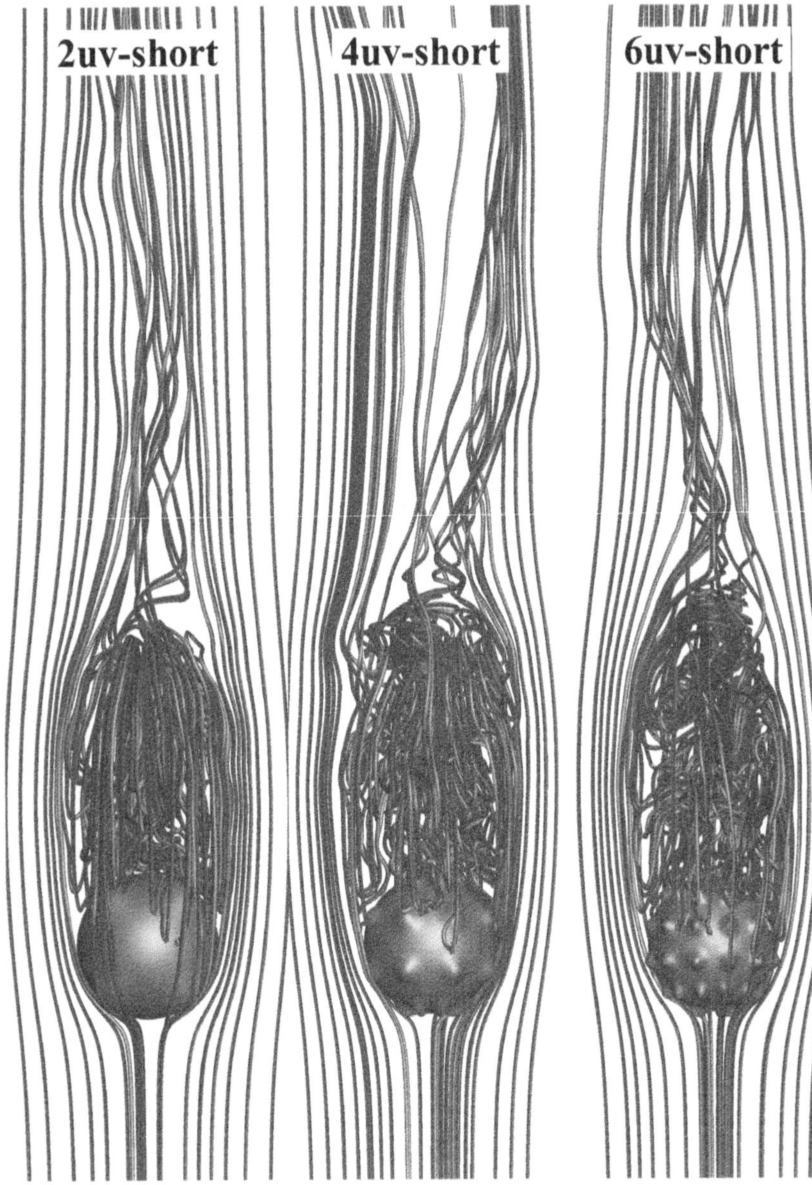

Fig. 4.75 A randomly selected snapshot of the streamtrace pattern of the computed flow field around three short lobed falling hailstones of 5 cm diameter ($N_{\mathrm{Re}} = 122{,}000$) (adapted from [53] with changes)

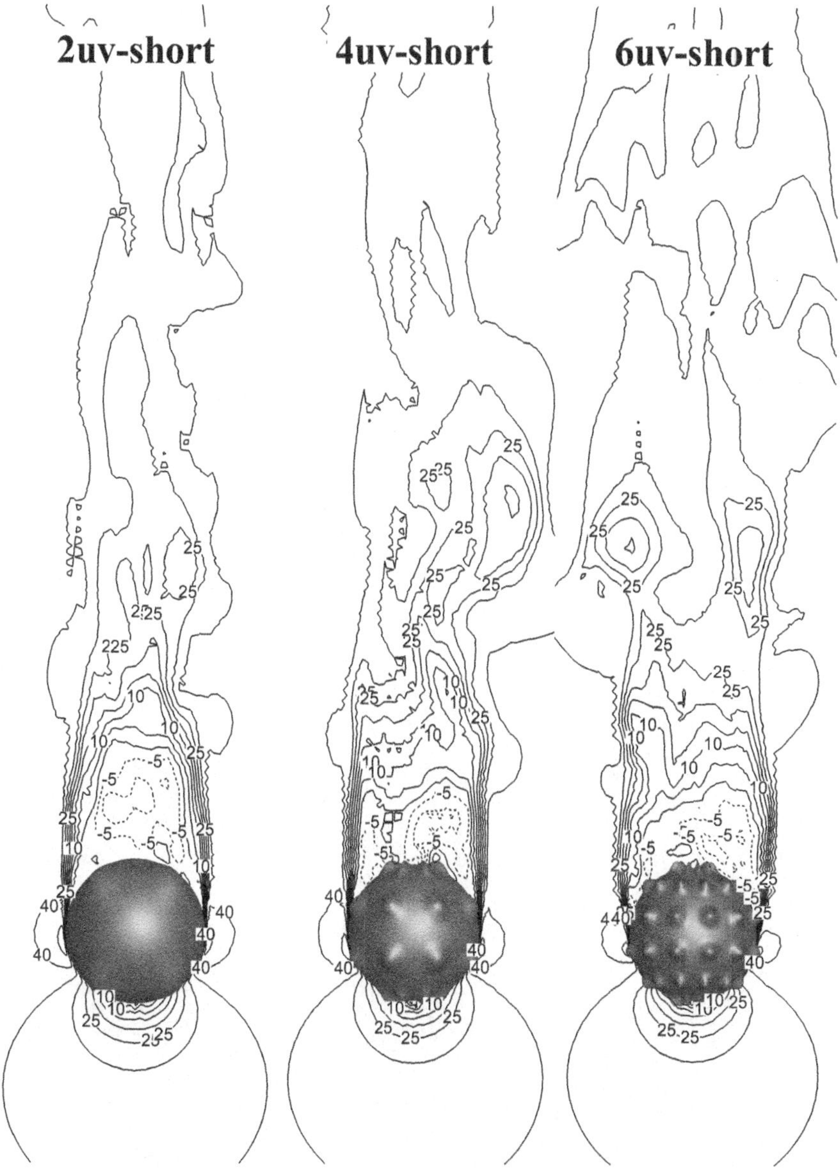

Fig. 4.76 The z-velocity field in the central yz-plane corresponding to Fig. 4.75 (adapted from [53] with changes)

Fig. 4.77 The pressure Field in the central YZ-plane corresponding to Fig. 4.75 (adapted from [53] with changes)

Figs. 4.75 and 4.76. It is true that the low pressure area becomes slightly smaller as the number of lobes increases. Figure 4.78 shows the corresponding vorticity magnitude fields of the three cases. Here we see that the main maximum vorticity regions still

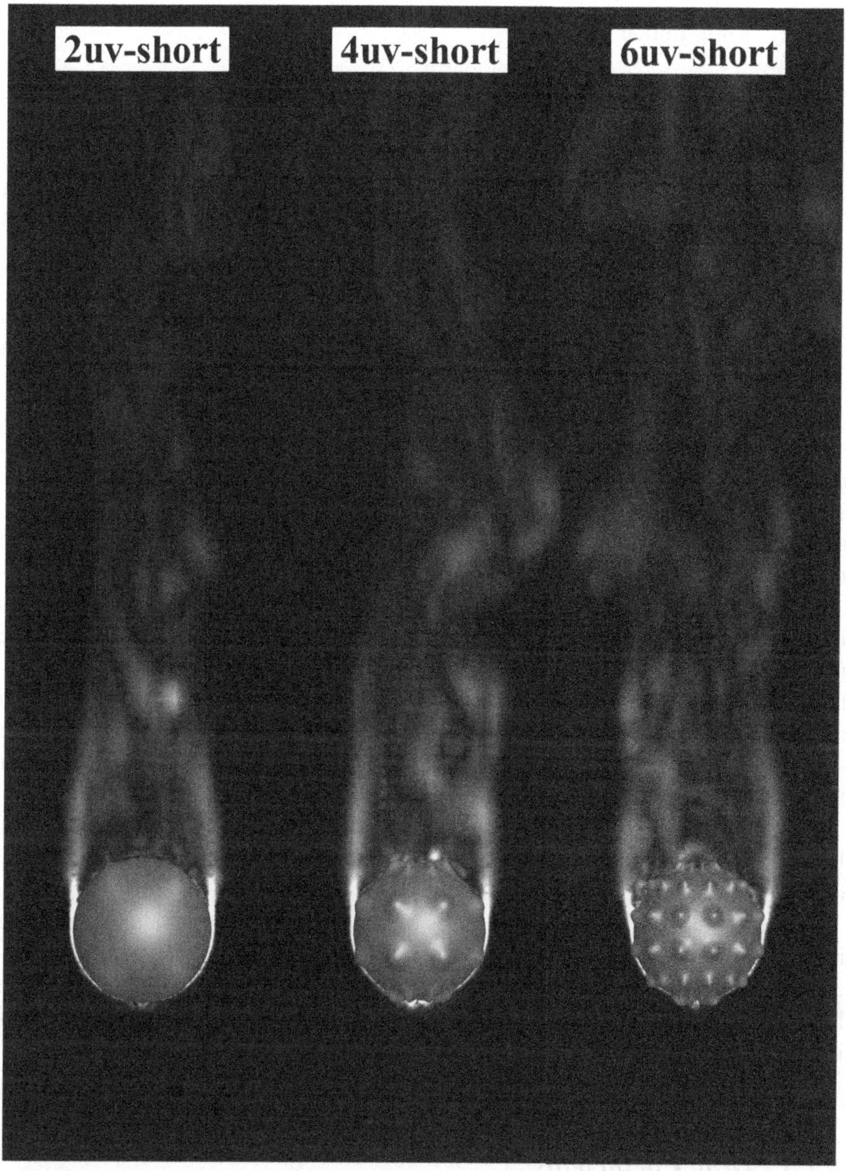

Fig. 4.78 The vorticity Field in the central YZ-plane corresponding to Fig. 4.75. The maximum and minimum vorticity magnitudes shown in these contours are 500 and 50,000 (1/s) (adapted from [53] with changes)

occur near the upstream of the equators of the hailstones, but their lengths of high vorticity magnitude become smaller compared with those of 1-cm case where the long stretched high vorticity flute flanking the recirculation bubble.

4.7.2.2 Long Lobed Cases

In order to understand the effects of how longer lobes on a hailstone impact the flow field and compare them with those of short lobed cases presented in the previous section, Figures 4.79, 4.80, 4.81, 4.82, 4.83, 4.84, 4.85 and 4.86 show the streamtrace, vertical velocity, and pressure distribution, vorticity magnitude fields, respectively, of the long lobe cases. Figures 4.79, 4.80, 4.81 and 4.82 are for the 1 cm and Figs. 4.83 and 4.86 are for the 5 cm cases.

The analysis of these figures allows us to draw the following conclusions:

The recirculation bubbles of 2uv, 4uv and 6uv long cases are close to each other in size in general but the eddies inside the recirculation bubbles are more chaotic as the number of lobes increases, similar to the short lobed cases in the previous section. As the Reynolds number and the number of lobes increase, the recirculation bubble sizes of long lobed hailstones decrease, again similar to the short lobed cases. Next, as the number of lobes goes up, the regions of the high downward speed become smaller. However, 2uv-long case has smaller regions of high downward speed even though it has fewer lobes. This result is not surprising since the hailstone has small number of and insignificantly long lobes.

The low pressure area behind the hailstone becomes larger as the number of lobes increases, similar to that for short lobed cases. As the Reynolds number and the number of lobes increase, the low pressure region behind the hailstone generally becomes smaller. The maximum vorticity magnitudes of both long and short lobed cases decrease as the number of lobes increases (see Fig. 4.87). This result can be explained by the more void space on the hailstone surface that can induce an earlier transition to turbulent regime, eventually leading to a reduced drag (see Fig. 4.88).

The flow patterns in Fig. 4.88 also seem to suggest the possibility that the longer lobes on these hailstones may affect the collection of aerosol particles or supercooled droplets by them.

The eddy lengths of most of the cases of the same diameter are almost the same regardless of the number of lobes, while 6uv long case has small size of its recirculation bubble in vertical length and in horizontal width. It is also clear that the horizontal width of the recirculation bubble becomes smaller as the number of lobes increases.

4.7.3 Drag Coefficients

The results of computed drag coefficients are shown in Figs. 4.89, 4.90, 4.91 and 4.92.

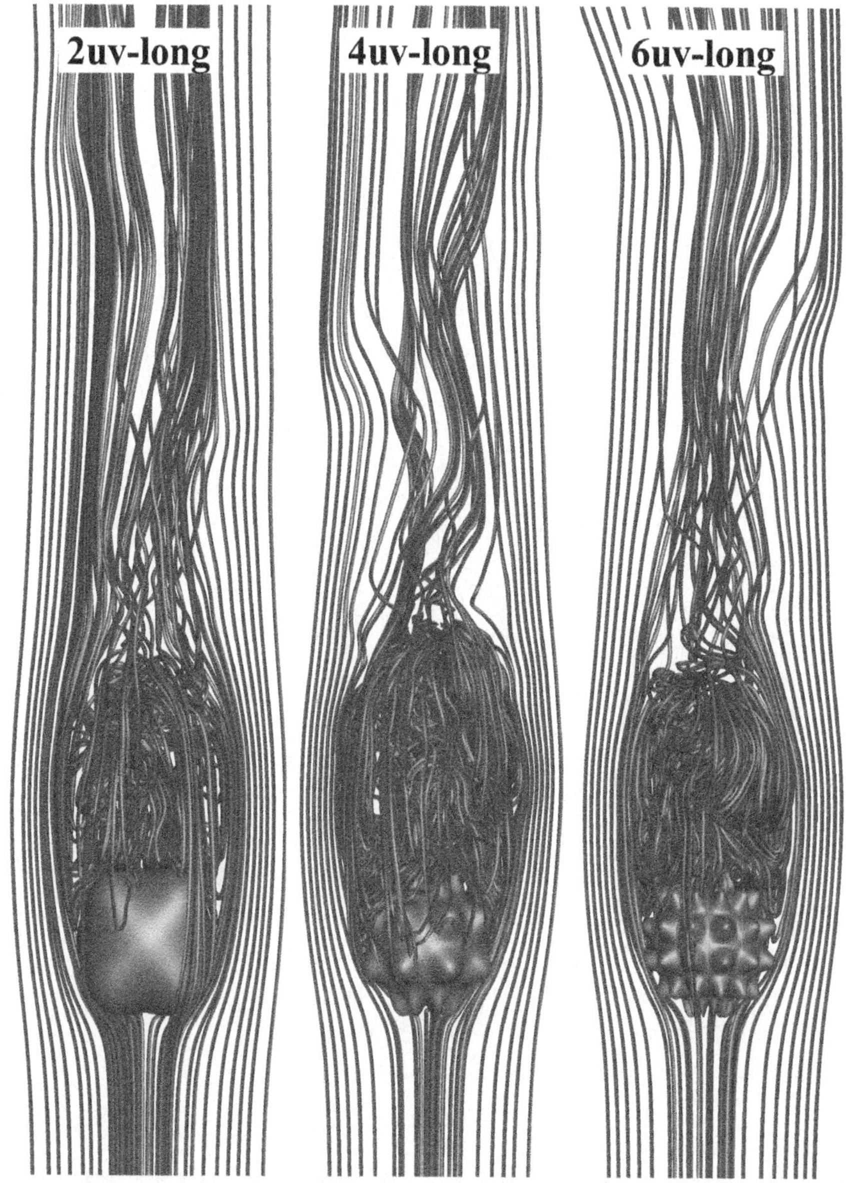

Fig. 4.79 A randomly selected snapshot of the streamtrace pattern of the computed flow field around three long lobed falling hailstones of 1 cm diameter ($N_{\mathrm{Re}} = 9560$) (adapted from [53] with changes)

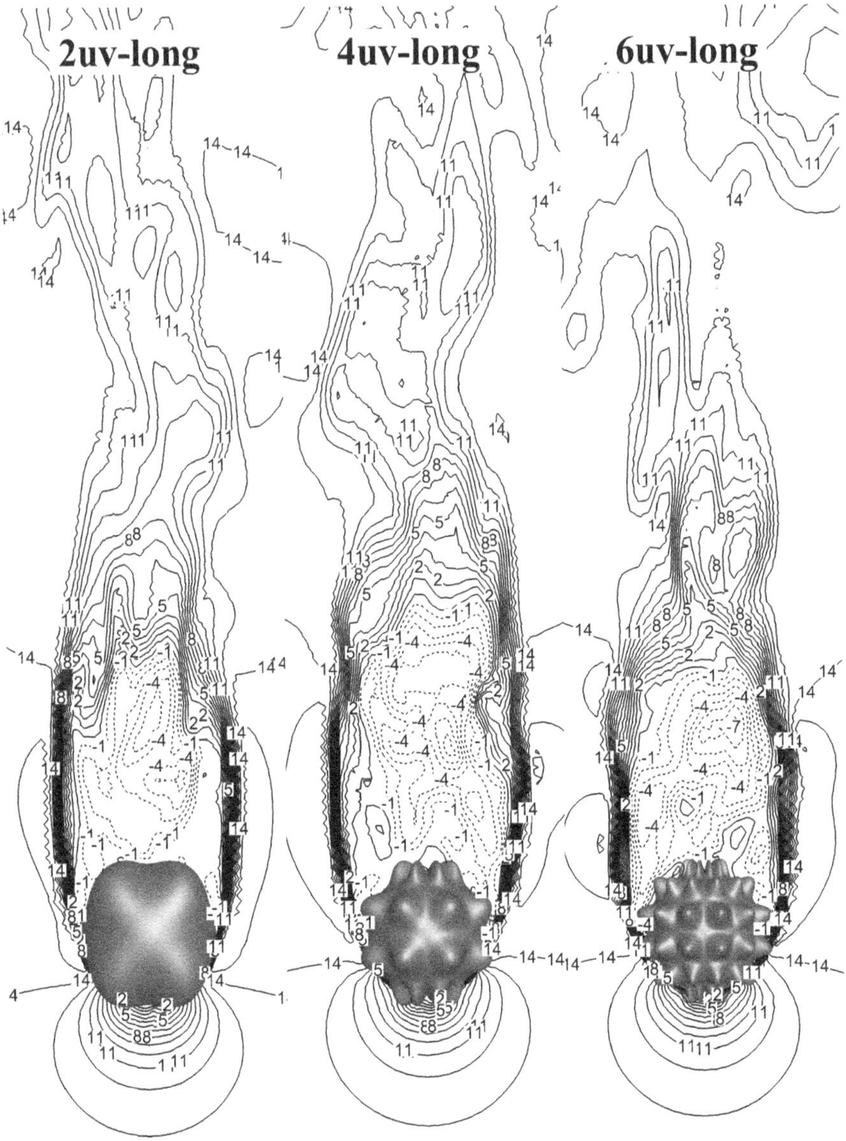

Fig. 4.80 The Z-velocity field in the central YZ-plane corresponding to Fig. 4.85 (adapted from [53] with changes)

The variation of drag coefficient with hail size (or Reynolds numbers) for different hailstones appear to be complicated. For both long and short lobed cases, the drag coefficients are higher for hailstones with more lobes. Among the long lobed hailstones (see Figs. 4.91 and 4.92), the drag coefficients of the 4uv and 6uv cases are

Fig. 4.81 The pressure field in the central YZ-plane corresponding to Fig. 4.79 (adapted from [53] with changes)

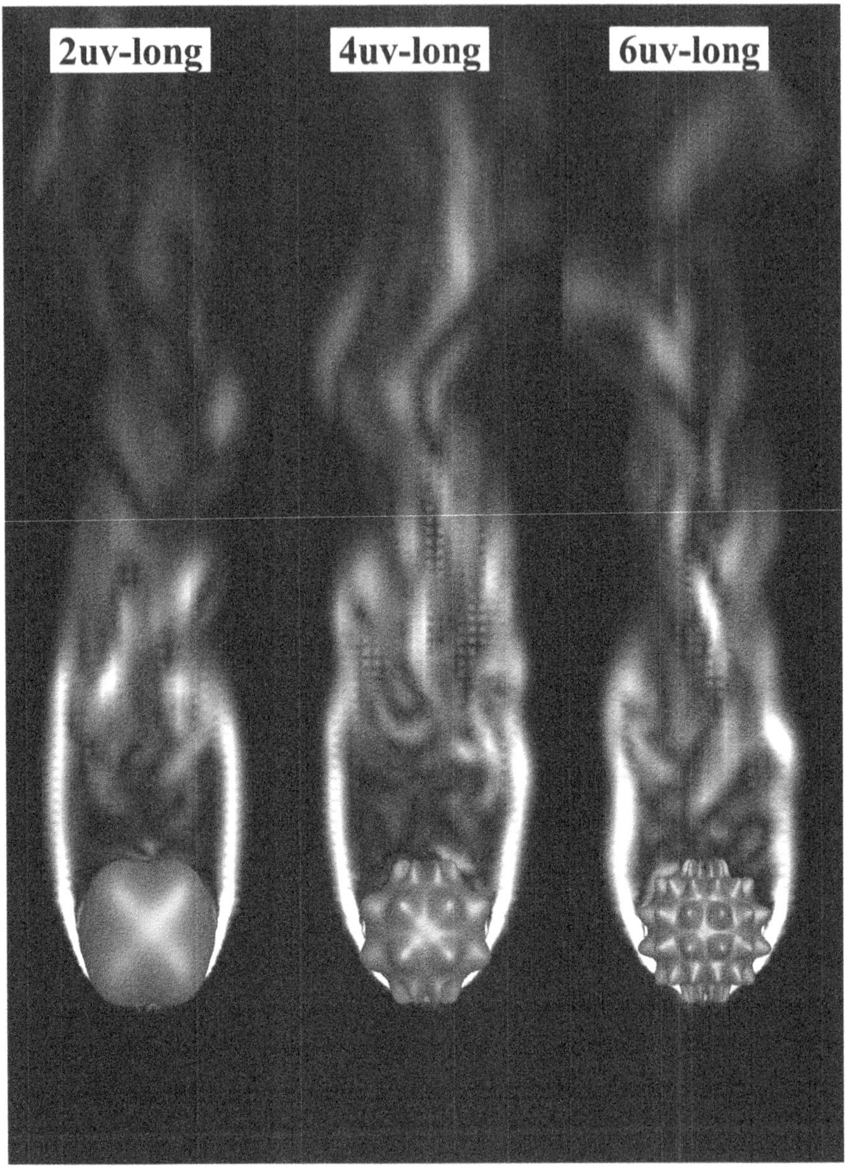

Fig. 4.82 The vorticity field in the central YZ-plane corresponding to Fig. 4.85. The maximum and minimum vorticity magnitudes shown in these contours are 500 and 50,000 (1/s) (adapted from [53] with changes)

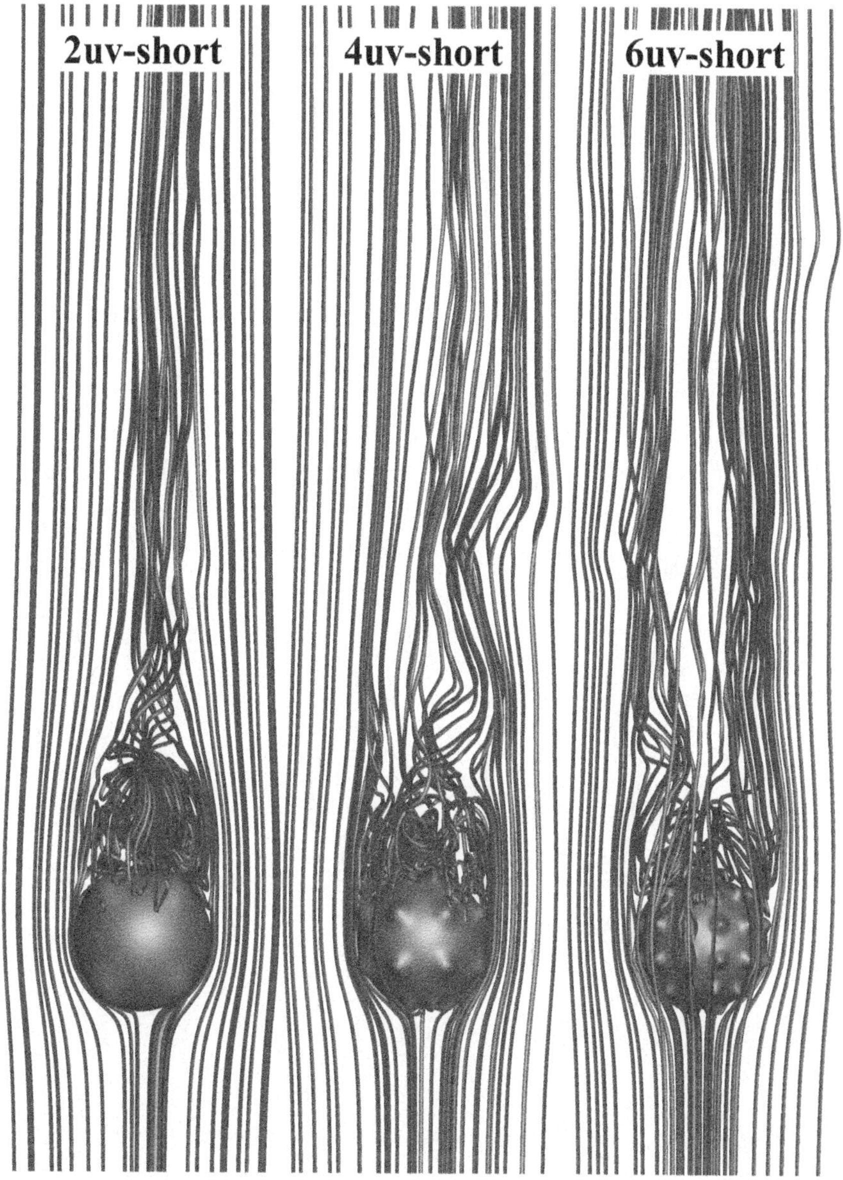

Fig. 4.83 A randomly selected snapshot of the streamtrace pattern of the computed flow field around three long lobed falling hailstones of 5 cm diameter ($N_{\text{Re}} = 122{,}000$) (adapted from [53] with changes)

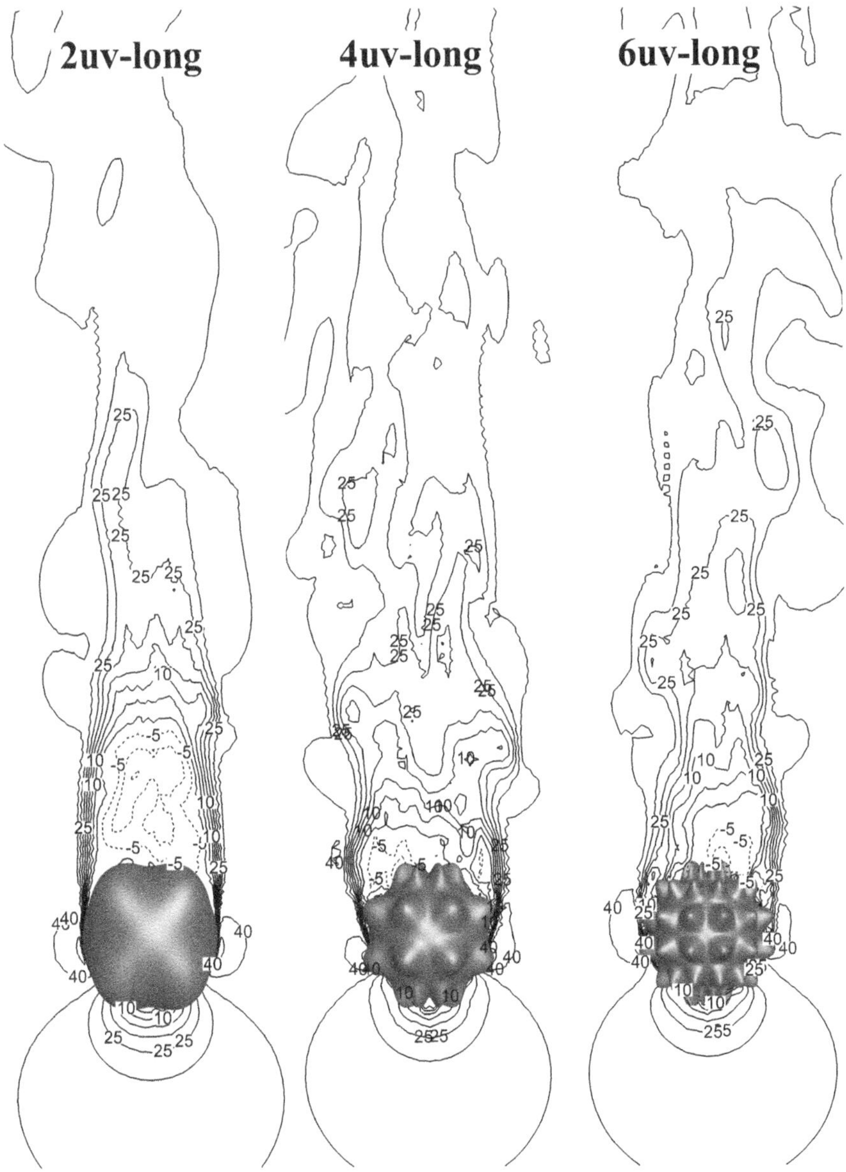

Fig. 4.84 The Z-velocity field in the central YZ-plane corresponding to Fig. 4.83 (adapted from [53] with changes)

Fig. 4.85 The pressure field in the central YZ-plane corresponding to Fig. 4.83 (adapted from [53] with changes)

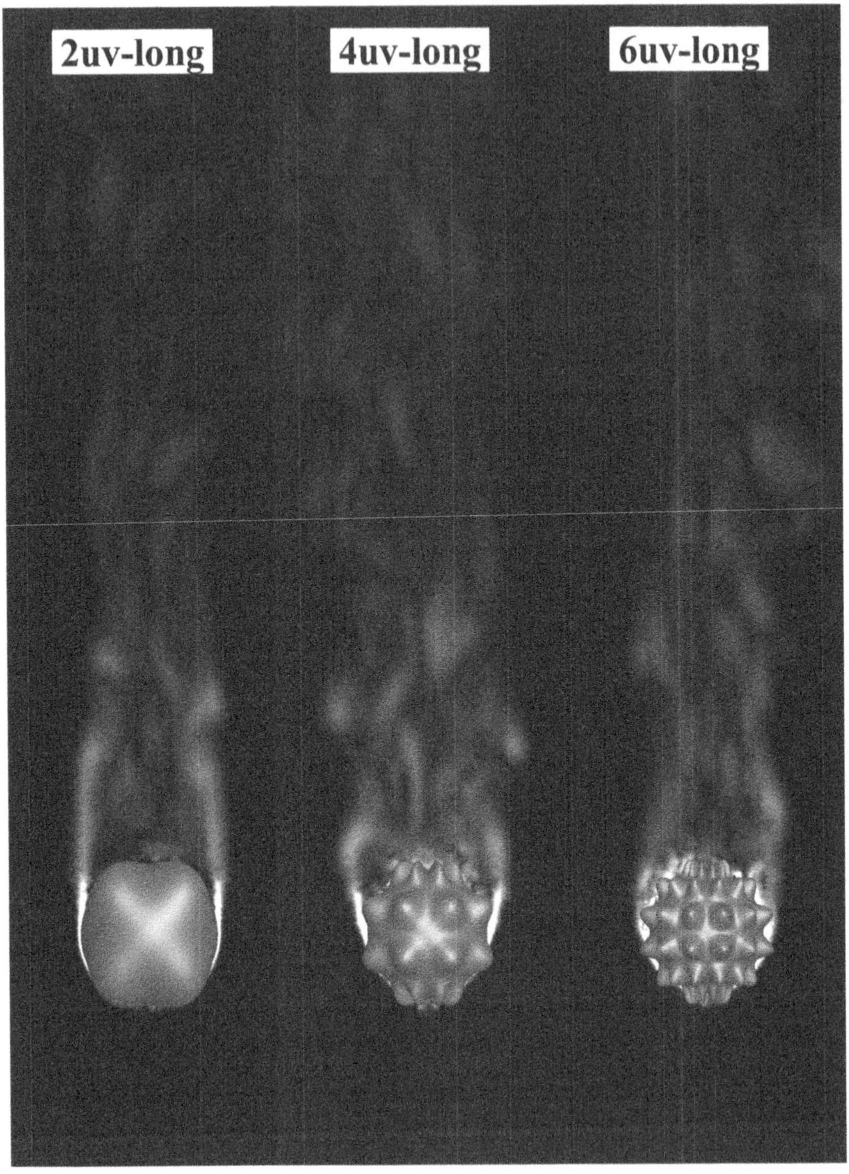

Fig. 4.86 The vorticity field in the central YZ-plane corresponding to Fig. 4.83. The maximum and minimum vorticity magnitudes shown in these contours are 500 and 50,000 (1/s) (adapted from [53] with changes)

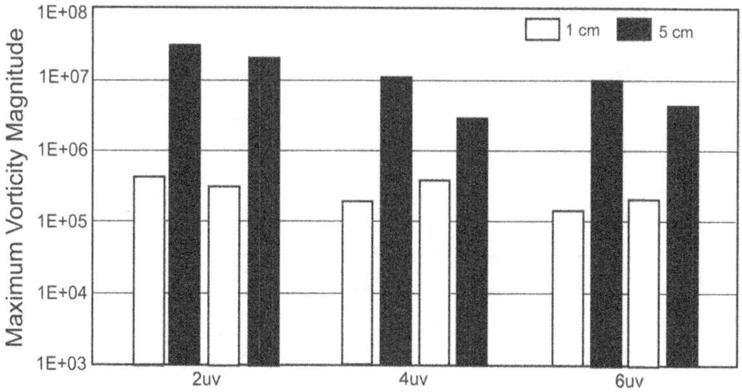

Fig. 4.87 Maximum Vorticity Magnitude of the long and short lobed hails of diameters 1 and 5 cm (adapted from [53] with changes)

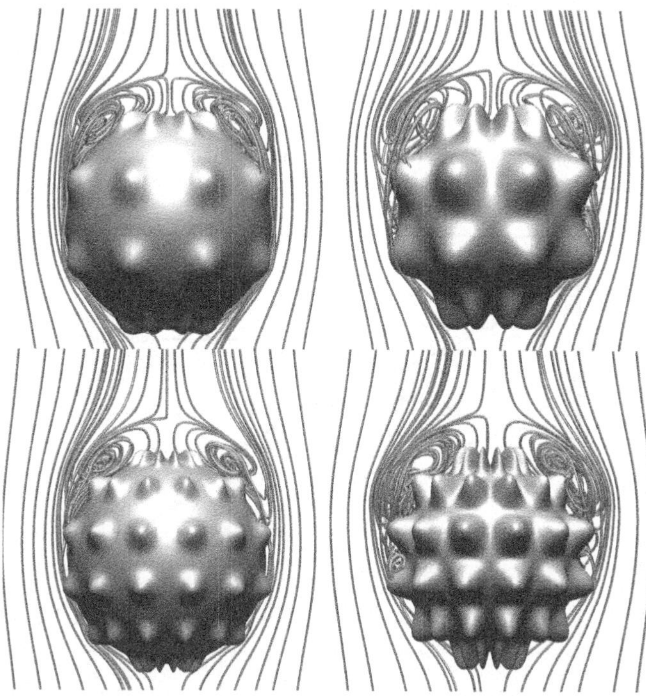

Fig. 4.88 The streamtrace patterns of 4uv and 6uv long and short lobed cases (adapted from [53] with changes)

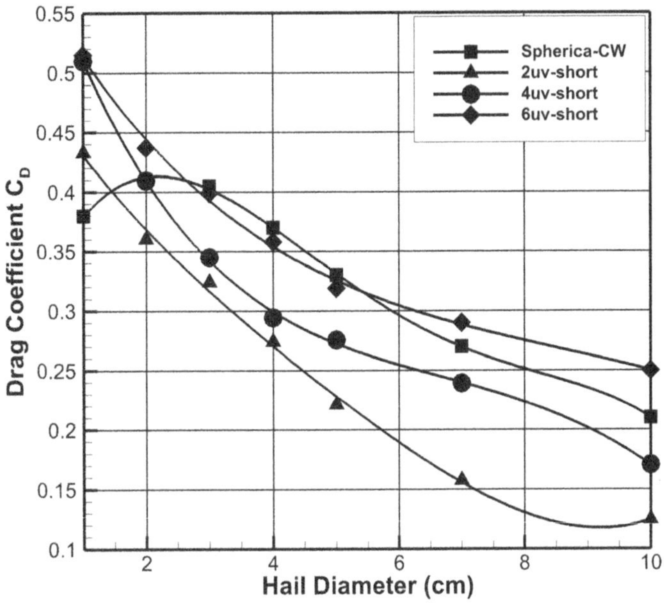

Fig. 4.89 Comparison of the drag coefficient C_D versus d plot for three short lobed hails using the previous empirical Formula (4.43) (adapted from [53] with changes)

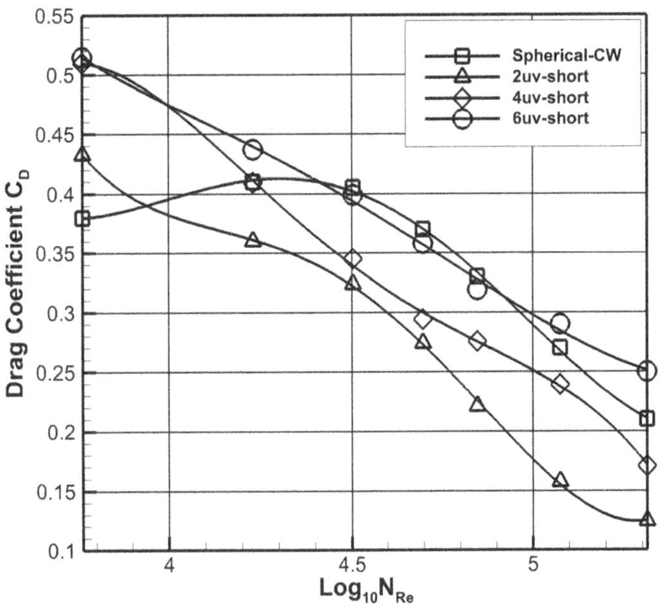

Fig. 4.90 Comparison of the drag coefficient C_D versus N_{Re} plot for three short lobed hails using the previous empirical Formula (4.44) (adapted from [53] with changes)

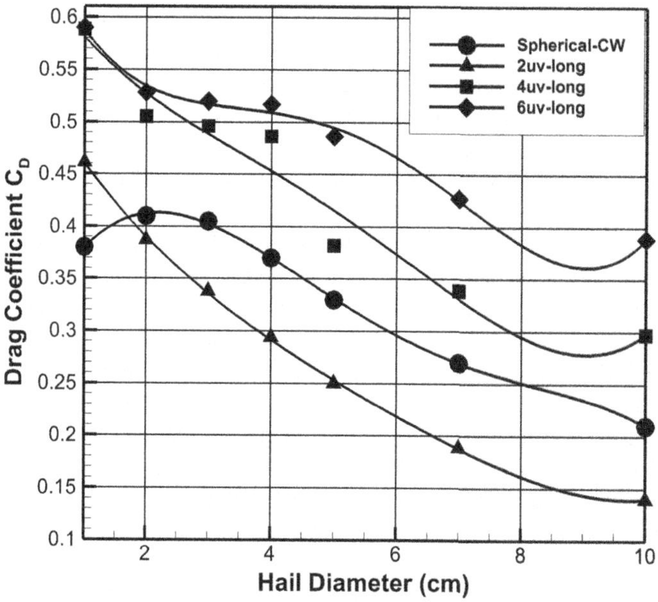

Fig. 4.91 Comparison of the drag coefficient C_D versus d plot for three long lobed hails using the previous empirical Formula (4.43) (adapted from [53] with changes)

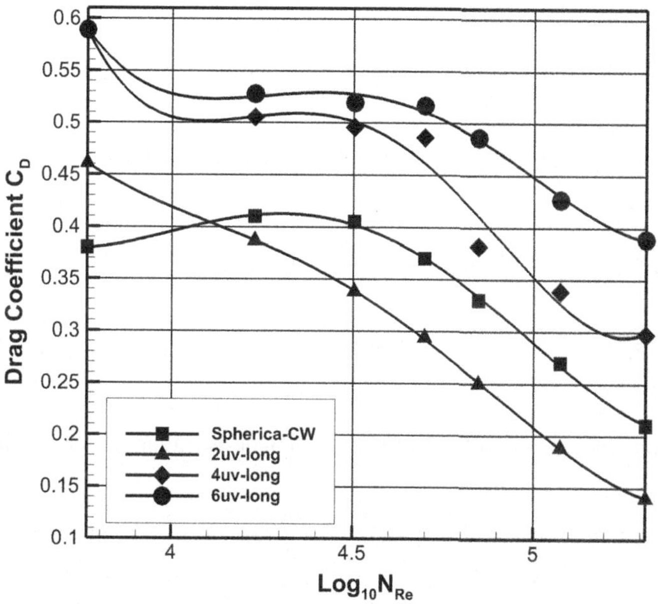

Fig. 4.92 Comparison of the drag coefficient C_D versus N_{Re} plot for three long lobed hails using the previous empirical Formula (4.44) (adapted from [53] with changes)

always higher than that of the spherical hailstones in Cheng and Wang [6] whereas most of the data points for 2uv case are lower than the spherical case.

On the other hand, for the short lobed hailstones (see Figs. 4.89 and 4.90), in the range of lower Reynolds numbers (roughly between zero and 25,000), 6uv and 4uv cases have higher drag coefficients than the spherical ones whereas 2uv cases have lower drag coefficients at the same Reynolds number. Thus, roughly speaking, in the Reynolds number range 25,000–122,000 all short lobed hailstone cases have lower drag coefficients compared to the corresponding spherical cases, which is consistent with previous observations that a sphere with suitably roughened surface may experience lower drag force than a smooth sphere of the same Reynolds number in this range. On the other hand, too rough a surface would not result in the decrease of drag as the long lobed cases here demonstrate. Thus we conclude that a short-lobed hailstone like the cases shown here will experience a smaller drag than a smooth spherical hailstone in this Reynolds number range whereas the drag of a long-lobed hailstone will experience a larger drag.

The computed values of drag coefficients as a function of diameter d can be fitted by Eqs. (4.41)–(4.46):

$$C_D = 4.1719 + 3.9005d - 3.3544d^2 + 5.8623d^3 - 2.8074d^4 \text{ (2uv - short)} \tag{4.41}$$

$$C_D = 6.2107 - 1.9508d + 3.7061d^2 - 3.2520d^3 + 1.0157d^4 \text{ (4uv - short)} \tag{4.42}$$

$$C_D = 5.5827 - 8.3902d + 4.4151d^2 + 3.7414d^3 - 3.2748d^4 \text{ (6uv - short)} \tag{4.43}$$

$$C_D = 4.7148 - 4.5711d - 1.2299d^2 + 2.6932d^3 - 1.3677d^4 \text{ (2uv - long)} \tag{4.44}$$

$$C_D = 6.3140 - 7.6981d - 2.4442d^2 + 1.7391d^3 - 1.1313d^4 \text{ (4uv - long)} \tag{4.45}$$

$$C_D = 5.6925 - 1.0563d - 6.7082d^2 + 7.3783d^3 - 1.8347d^4 \text{ (6uv - long)} \tag{4.46}$$

Equations (4.47)–(4.52) fit the drag coefficient as a function of X where $X = \log_{10} N_{\text{Re}}$:

$$C_D = -6.5530 + 5.2674X - 1.5534X^2 + 2.0051X^3 - 9.5928X^4 \text{ (2uv - short)} \tag{4.47}$$

$$C_D = -1.0069 + 8.6704X - 2.7610X^2 + 3.8747X^3 - 2.0268X^4 \text{ (4uv - short)} \tag{4.48}$$

$$C_D = -1.5411 + 1.2215X - 3.3773X^2 + 3.9567X^3 - 1.6638X^4 \text{ (6uv - short)} \quad (4.49)$$

$$C_D = -2.5460 + 1.9734X - 5.4543X^2 + 6.4433X^3 - 2.7531X^4 \text{ (2uv - long)} \quad (4.50)$$

$$C_D = -8.6501 + 7.3407X - 2.3015X^2 + 3.1841X^3 - 1.6450X^4 \text{ (4uv - long)} \quad (4.51)$$

$$C_D = 1.3470 - 1.1453X + 3.6493X^2 - 5.1400X^3 + 2.6973X^4 \text{ (6uv - long)} \quad (4.52)$$

It is clear that a hailstone with long or more lobes has a higher drag coefficient than one with short or fewer lobes.

In the above, we have presented the numerically solved flow fields around the falling lobed hailstones of 7 different diameters (1–10 cm). Their shapes are simulated by the formulas based on a method described by Wang [51, 52]. The characteristics of the simulated flow fields are consistent with available experiments obtained by previous investigators. This demonstrates that the CFD simulations are useful for numerically predicting the realistic flow fields around the falling lobed hailstones.

It is often difficult to compare numerical calculations of flow field results with direct field observations regarding graupel and hailstones as uncertainties in field results can be large. Laboratory experiments such as those conducted in a vertical wind tunnel, on the other hand, can provide more precise data for such comparison as the environmental conditions can be controlled more precisely. Recent measurements by Heymsfield et al. [16] and Jost et al. [23] conducted in the Mainz Wind Tunnel have provided more definite results and the numerical results generally compare well with the experimental data.

References

1. Anderson JD (2004) Introduction to flight, 5th edn. McGraw-Hill, 616 pp
2. Auer AH, Veal DL (1970) The dimension of ice crystals in natural clouds. J Atmos Sci 27:919–926
3. Belmonte A, Eisenberg H, Moses E (1998) From flutter to tumble: inertial drag and froude similarity in falling paper. Phys Rev Lett 81:345–348. https://doi.org/10.1103/PhysRevLett.81.345
4. Böhm HP (1989) A general equation for the terminal fall speed of solid hydrometeors. J Atmos Sci 46:2419–2427
5. Böhm HP (1992) A general hydrodynamic theory for mixed-phase microphysics. Part I: drag and fall speed of hydrometeors. Atmos Res 27:253–274
6. Cheng K-Y, Wang PK (2013) A numerical study of the flow fields around falling hails. Atmos Res 132–133:253–263
7. Cheng KY, Wang PK, Hashino T (2015) A numerical study on the attitudes and aerodynamics of freely falling hexagonal ice plates. J Atmos Sci 72:3685–3698

8. Chueh CC, Wang PK, Hashino T (2017) A preliminary numerical study on the time-varying fall attitudes and aerodynamics of freely falling conical graupel particles. Atmos Res 183:58–72
9. Chueh C-C, Wang PK, Hashino T (2018) Numerical study of motion of falling conical graupel. Atmos Res 199:82–92
10. Constantinescu G, Squires K (2004) Numerical investigations of flow over a sphere in the subcritical and supercritical regimes. Phys Fluids 16:1449–1466
11. Cox RG (1965) The steady motion of a particle of arbitrary shape at small Reynolds numbers. J Fluid Mech 23:625–643
12. Gravner J, Griffeath D (2009) Modeling snow-crystal growth: a three-dimensional mesoscopic approach. Phys Rev E 79:011601
13. Hashino T, Cheng K-Y, Chueh C-C, Wang PK (2016) Numerical study of motion and stability of falling columnar crystals. J Atmos Sci 73:1923–1942
14. Hashino T, Chiruta M, Polzin D, Kubicek A, Wang PK (2014) Numerical simulation of the flow fields around falling ice crystals with inclined orientation and the hydrodynamic torque. Atmos Res 150:79–96
15. Heymsfield AJ, Kajikawa M (1987) An improved approach to calculating terminal velocities of plate-like crystals and graupel. J Atmos Sci 44:1088–1099. https://doi.org/10.1175/1520-0469(1987)044%3c1088:AIATCT%3e2.0.CO;2
16. Heymsfield A, Szakáll M, Jost A, Giammanco I, Wright R (2018) A comprehensive observational study of graupel and hail terminal velocity, mass flux, and kinetic energy. J Atmos Sci 75(11):3861–3885
17. Heymsfield AJ, Westbrook CD (2010) Advances in the estimation of ice particle fall speeds using laboratory and field measurements. J Atmos Sci 67:2469–2482
18. Ishimoto H (2008) Radar backscattering computations for fractal-shaped snowflakes. J Meteor Soc Jpn Ser II 86:459–469. https://doi.org/10.2151/jmsj.86.459
19. Jayaweera KOLF, Mason BJ (1965) The behavior of freely falling cylinders and cones in a viscous fluid. J Fluid Mech 22:709–720
20. Ji W, Wang PK (1989) Numerical simulation of three dimensional unsteady viscous flow past hexagonal ice crystals in the air—preliminary results. Atmos Res 25:539–557
21. Ji W, Wang PK (1990) Numerical simulation of three-dimensional unsteady viscous flow past fixed hexagonal ice crystals in the air—preliminary results. Atmos Res 25:539–557. https://doi.org/10.1016/0169-8095(90)90037-D
22. Ji W, Wang PK (1991) Numerical simulation of three-dimensional unsteady viscous flow past finite cylinders in an unbounded fluid at low intermediate Reynolds numbers. Theor Comput Fluid Dyn 3:43–59
23. Jost A, Szakáll M, Diehl K, Mitra SK, Hundertmark A, Klug BS, Borrmann S (2019) The effect of turbulence on the accretional growth of graupel. J Atmos Sci 76:3047–3061
24. Kajikawa M (1976) Observation of falling motion of columnar snow crystals. J Meteor Soc Jpn 54:276–284
25. Kajikawa M (1992) Observations of the falling motion of plate-like snow crystals. Part I: the free-fall patterns and velocity variations of unrimed crystals. J Meteor Soc Jpn 70:1–9
26. Knight NC, Heymsfield AE (1983) Measurement and interpretation of hailstone density and terminal velocity. J Atmos Sci 40:1510–1516
27. Kubicek A, Wang PK (2012) A numerical study of the flow fields around a typical conical graupel falling at various inclination angles. Atmos Res 118:15–26
28. List R, Schemenauer RS (1971) Free-fall behavior of planar snow crystals, conical graupel and small hail. J Atmos Sci 28:110–115. https://doi.org/10.1175/1520-0469(1971)028%3c0110:FFBOPS%3e2.0.CO;2
29. Locatelli JD, Hobbs PV (1974) Fall speeds and masses of solid precipitation particles. J Geophys Res 79:2185–2197. https://doi.org/10.1029/JC079i015p02185
30. Massey BS (1979) Mechanics of fluids. Van Nostrand Reinhold, 543 pp
31. Matrosov SY (2007) Modeling backscatter properties of snowfall at millimeter wavelengths. J Atmos Sci 64:1727–1736. https://doi.org/10.1175/JAS3904.1

32. Mitchell DL (1996) Use of mass- and area-dimensional power laws for determining precipitation particle terminal velocities. J Atmos Sci 53:1710–1723
33. Mitchell DL, Heymsfield AJ (2005) Refinements in the treatment of ice particle terminal velocities, highlighting aggregates. J Atmos Sci 62:1637–1644
34. Mitchell DL, Zhang R, Pitter RL (1990) Mass-dimensional relationships for ice particles and the influence of riming on snowfall rates. J Appl Meteor 29:153–163
35. Nettesheim J, Wang PK (2018) A numerical study on the aerodynamics of freely falling planar ice crystals. J Atmos Sci 75:2849–2865
36. Pflaum JC, Martin JJ, Pruppacher HR (1978) A wind tunnel investigation of the hydrodynamic behaviour of growing, freely falling graupel. Q J R Meteor Soc 104:179–187
37. Pitter RL, Pruppacher HR (1974) A numerical investigation of collision efficiencies of simple ice plates colliding with supercooled water drops. J Atmos Sci 31:551–559
38. Pitter RL, Pruppacher HR, Hamielec AE (1973) A numerical study of viscous flow past a thin oblate spheroid at low and intermediate Reynolds numbers. J Atmos Sci 30:125–134
39. Platt CMR (1978) Lidar backscatter from horizontal ice crystal plates. J Appl Meteor 17:482–488
40. Pope S (2000) Turbulent flows. Cambridge University Press, 802 pp
41. Pruppacher HR, Klett JD (1997) Microphysics of clouds and precipitation. 2nd edn. D. Reidel, New York, 954 pp
42. Rodriguez I, Lehmkuhl O, Borrell R, Oliva A, Perez-Segarra CD (2010) Direct numerical simulation of turbulent wakes: flow past a sphere at Re = 5000. In: Pereira JCF, Sequeira A, Pereira JMC (eds) Proceedings of the V European conference on computational fluid dynamics. ECCOMAS CFD 2010, Lisbon, 14–17 June 2010. ISBN 978-989-96778-1-4
43. Roos FW, Willmarth WW (1971) Some experimental results on sphere and disk drag. AIAA J 9:285–291. https://doi.org/10.2514/3.6164
44. Schlamp RJ, Pruppacher HR, Hamielec AE (1975) A numerical investigation of the efficiency with which simple columnar ice crystals collide with supercooled water drops. J Atmos Sci 32:2330–2337
45. Sen S, Mittal S, Biswas G (2009) Steady separated flow past a circular cylinder at low Reynolds numbers. J Fluid Mech 620:89–119
46. Stringham GE, Simons DB, Guy HP (1969) The behavior of large particles falling in quiescent liquids. US Government Printing Office, 36 pp
47. Tomboulides AG, Orszag SA (2000) Numerical investigation of transitional and weak turbulent flow past a sphere. J Fluid Mech 416:45–73
48. Van Dyke M (1982) An album of fluid motion. The Parabolic Press, Stanford, 176 pp
49. Wang PK (1982) Mathematical description of the shape of conical hydrometeors. J Atmos Sci 39:2615–2622
50. Wang PK (1983) On the definition of collision efficiency of atmospheric particles. J Atmos Sci 40:1051–1052
51. Wang PK (1997) Characterization of ice crystals in clouds by simple mathematical expressions based on successive modification of simple shapes. J Atmos Sci 54:2035–2041
52. Wang PK (1999) Three-dimensional representations of hexagonal ice crystals and hailstone particles of elliptical cross sections. J Atmos Sci 56:1089–1093
53. Wang PK, Chueh CC, Wang CK (2015) A numerical study of flow fields of lobed hailstones falling in air. Atmos Res 160:1–14
54. Wang PK, Ji W (1997) Numerical simulation of three-dimensional unsteady flow past ice crystals. J Atmos Sci 54:2261–2274
55. Wang PK, Kubicek A (2013) Flow fields of graupel falling in air. Atmos Res 124:158–169
56. Wang PK, Pruppacher HR (1977) An experimental determination of the efficiency with which aerosol particles are collected by water drops in subsaturated air. J Atmos Sci 34:1664–1669
57. Wang PK (2013) Physics and dynamics of clouds and precipitation. Cambridge University Press, 467 pp
58. Willmarth WW, Hawk NE, Harvey RL (1964) Steady and unsteady motions and wakes of freely falling disks. Phys Fluids 7:197. https://doi.org/10.1063/1.1711133
59. Zikmunda J, Vali G (1972) Fall patterns and fall velocities of rimed ice crystals. J Atmos Sci 29:1334–1347

Chapter 5
Ventilation Effect of Falling Ice Hydrometeors

5.1 Introduction

In Chap. 4, we described the fall behavior of ice hydrometeors and showed that different hydrometeors induced very different flow fields around themselves during the fall. One of the main purposes of studying these flow fields is to understand how these fields influence the growth rates of ice hydrometeors. The growth (and its opposite process, dissipation) rates of hydrometeors in a cloud determine how fast this cloud can develop and how long it can last in the atmosphere.

There are generally two major growth modes of hydrometeors: collision growth and diffusion growth [13, 17]. The former involves the collision and subsequent coalescence of two or more particles whereas the latter involves the diffusion of water vapor towards the surface of a hydrometeor when the environment is supersaturated. It is the latter that we want to discuss more in this paper.

As we mentioned in Chap. 3, when hydrometeors are falling relative to air, the water vapor density distribution around the falling particle will be influenced by the flow field around the particle due to the motion. Thus for a stationary spherical hydrometeor, the vapor distribution should be spherically symmetric, that is, it depends only on the radial coordinate r. For a falling spherical hydrometeor, however, the vapor density field would be spherically asymmetric and the details of the asymmetry depend on the flow. In general, the vapor density gradients will be enhanced in the upstream portion of the hydrometeor while that in the rear would be relaxed when compared to that of a stationary spherical hydrometeor. When averaged over the entire surface of the hydrometeor, the gradient (in terms of magnitude) is always greater than that around a stationary one. Thus there is an overall enhancement of the vapor density gradient. This enhancement factor is called the *ventilation coefficient* (see, e.g., [17]). When a hydrometeor is falling in subsaturated air, it will evaporate faster than a stationary one by the same factor, i.e., the ventilation coefficient.

Both diffusion growth and evaporation involve phase change and hence are associated with the release and consumption of latent heat respectively. The heat transfer problem during the fall of the hydrometeor is analogous to the vapor diffusion, i.e.,

© Springer Nature Singapore Pte Ltd. 2021

P. K. Wang, *Motions of Ice Hydrometeors in the Atmosphere*, Atmosphere, Earth, Ocean & Space, https://doi.org/10.1007/978-981-33-4431-0_5

it is usually considered as a heat diffusion process. Thus the heat diffusion rate will be enhanced by the motion and there is also a ventilation coefficient relevant for it. In the context of cloud physics, it is often assumed that the ventilation coefficient for heat diffusion is the same as that for vapor diffusion [13].

Ventilation coefficients for some falling hydrometeors have been investigated before both experimentally and theoretically. For example, Beard and Pruppacher [3] performed experiments to measure the ventilation coefficients for small evaporating water droplets falling in air utilizing a vertical wind tunnel in UCLA. Woo and Hamielec [19] numerically calculated the ventilation coefficients for small water droplets. Pruppacher and Rasmussen [14] performed similar but more accurate measurements for large drops again using the UCLA vertical wind tunnel.

For small ice crystals, Thorpe and Mason [16] measured experimentally the ventilation coefficients for falling hexagonal ice crystals. Masliyah and Epstein [10] and Pitter et al. [12] performed numerical calculations of ventilation coefficients using thin oblate spheroid to represent plate ice crystals. Ji and Wang [7] numerically computed the ventilation coefficients for falling ice crystals using more realistic shapes of ice columns, hexagonal plates and broad branch ice crystals.

For very large ice particles, especially hailstones, List [8] studied the general heat and mass transfer during the accretion growth of large hail whereas Macklin [9] measured the heat transfer rate of melting ice spheres and spheroids of diameter or long axis of 3.8 and 5.1 cm. Bailey and Macklin [2] repeated similar experiments for spheroids with diameters 3.0–7.2 cm. From these heat transfer rates one can also derive the ventilation coefficients although they are confined to the accretion growth or melting mode of hail.

There appears to be no systematic theoretical studies to determine the ventilation coefficients for falling large ice hydrometeors. The main reason is the same as that for theoretical study of the flow fields around large falling ice particles—the limitation of computer technology and advanced software, both become available now. This motivates the current study.

The equations involved in the calculations of the ventilation coefficients of falling ice hydrometeors have been given in Sect. 3.3. For the physical setting of the numerical simulations, we assume that the environmental relative humidity (RH) far away from the crystal is 102% while RH = 100% at the ice hydrometeor surface. In other words, we are assuming that the crystal is in a growth mode. Note that the RH we are considering here is the RH with respect to ice surface which is generally higher than that with respect to liquid water surface at the same temperature for temperature range 0 to −40 °C (see [17, p. 100]). Note, however, that the ventilation coefficient so calculated will be independent of the assumed humidity condition as it is purely a function of the ice particle motion. The same ventilation coefficient will apply to both growth and evaporation modes for the same particle under the same condition of motion.

In the following, we will show the calculation results and their implications.

5.2 Vapor Density Distributions and Ventilation Coefficients

5.2.1 Planar Ice Crystals

We will use the cases of hexagonal ice plates, conical graupel and smooth and lobed hailstones to illustrate the vapor distribution and the ventilation phenomenon. Figure 5.1 shows the vapor density distribution in the central vertical x–z cross-section (y = 0) around a falling hexagonal ice plate of d = 2 mm. As we have described in Chap. 4, a 2 mm hexagonal ice plate would fall steadily and the flow field an axisymmetric type. Consequently, one would expect that the distribution of water vapor density should also be axisymmetric and Fig. 5.1 shows that this is indeed the case. But of course there can be no fore-and-aft symmetry as the plate is falling. As we explained before, the motion would force the vapor to become more concentrated (lighter shade) in the lower (upstream) part and more dilute (darker shade) in the upper (downstream) part of the ice plate than when the plate is stationary. This is manifested by the crowded vapor isopycnals in the lower side and relaxed isopycnals in the upper side of the plate. The regions with the least crowded isopycnals are on the both side of plate in the downstream, regions that correspond to high air velocities upward (see Fig. 4.7) that carry water vapor quickly away from the plate surface.

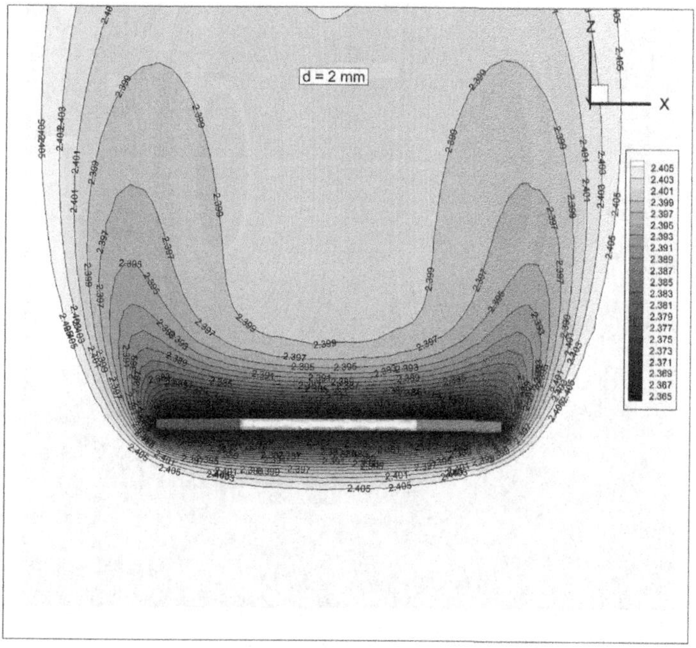

Fig. 5.1 Vapor density distribution around a felling hexagonal ice plate of $d = 2$ mm

In 3-D sense, this "void" of water vapor would form a cup shape. In comparison, the central downstream wake region is a relative high density region because the air velocities are downward-pointing in the recirculation bubble and carry the vapor towards the ice surface.

Figure 5.2 shows the **vapor density gradient** distribution corresponding to Fig. 5.1. It shows that the highest vapor density gradient occurs at the underside of the vertices of the hexagon whereas the minimum gradient occurs just outside of the low vapor density cup shown in Fig. 5.1. Figure 5.3 shows another view that demonstrates even better this distribution. The orange color contour surfaces show the highest vapor density gradient sites which are right on the underside of the six vertices. Since the mass diffusion flux is proportional to the vapor density gradient as indicated in Eq. (3.15), the linear growth rate of the plate will be the fastest wherever the vapor density gradient is the largest. This means that the vertices are where the fastest growth location one can expect from diffusion process. Judging from the comprehensive ice crystal photographs catalogued by Bentley and Humphreys [4] that the new branches of simple ice crystals with a hexagonal plate in the center all stem from the vertices, it appears that the vapor gradient indeed plays the most important role in guiding the growth direction and hence determining the ice habit.

The above observation only applies to the case of steady fall where the crystal falls vertically so that the vapor gradient distribution is hexagonally symmetric. This usually occurs when the ice crystals are relatively small. When crystals become

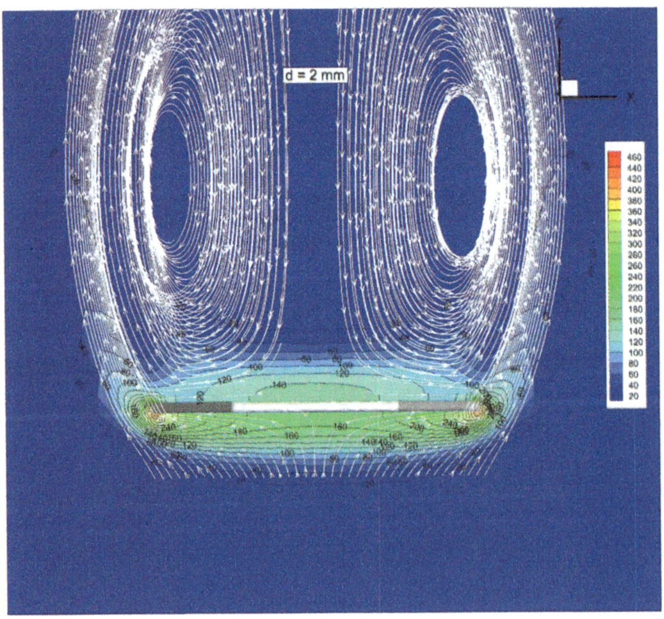

Fig. 5.2 Vapor density gradient field around a falling hexagonal ice crystal of $d = 2$ mm. White contours are streamtraces

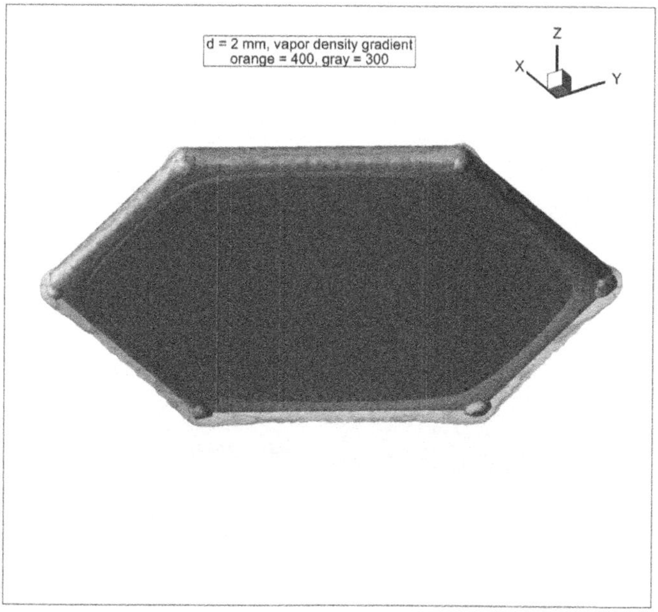

Fig. 5.3 Isosurfaces of vapor density gradients 400 (orange) and 300 (grey) around a falling hexagonal ice plate of $d = 2$ mm

larger, they tend to fall with unsteady motions as illustrated in Chap. 4 and such symmetry will be broken.

Next, we will examine the case of a larger hexagonal ice plate that has a diameter of 5 mm. Figure 5.4 shows a randomly selected x–z cross-sectional frame of the distribution of water vapor around this plate. A 5 mm diameter hexagonal ice plate no longer falls steadily but rather perform unsteady zigzag oscillation and rotation during the fall. Thus the vapor distribution is no longer axisymmetric as is clear from this figure. Other than this, the general features of the vapor density distribution remain similar to that in Fig. 5.1.

Figure 5.5 shows the vapor density gradient distribution in the x–z cross-section. Unlike Fig. 5.2 which shows the hexagonal symmetric distribution with the maxima at the vertices, the maximum vapor density gradient now is located at the lowest point of the plate due to the non-axisymmetric flow field. This feature is made even clearer by Fig. 5.6 which shows the 3-D vapor density gradient distribution. From the growth rate comment we mentioned before, the fastest growth will occur at the point of maximum vapor density gradient, i.e., at the lowest point of the plate. Now, due to the oscillatory motion of the plate of this size and the oscillation tends to occur on a preferential plane, this will result in a asymmetric growth of the branches, some new branches will grow longer than other branches which are indeed observed in many natural snow crystals.

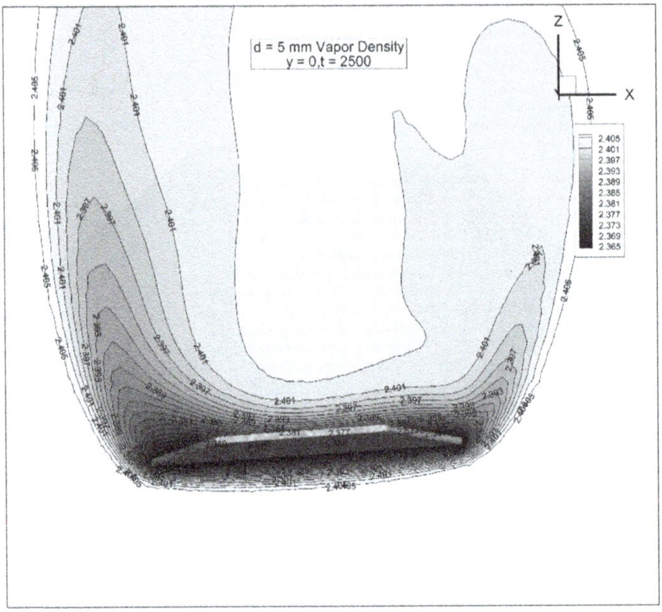

Fig. 5.4 A randomly selected snapshot of the vapor density distribution around a felling hexagonal ice plate of $d = 5$ mm

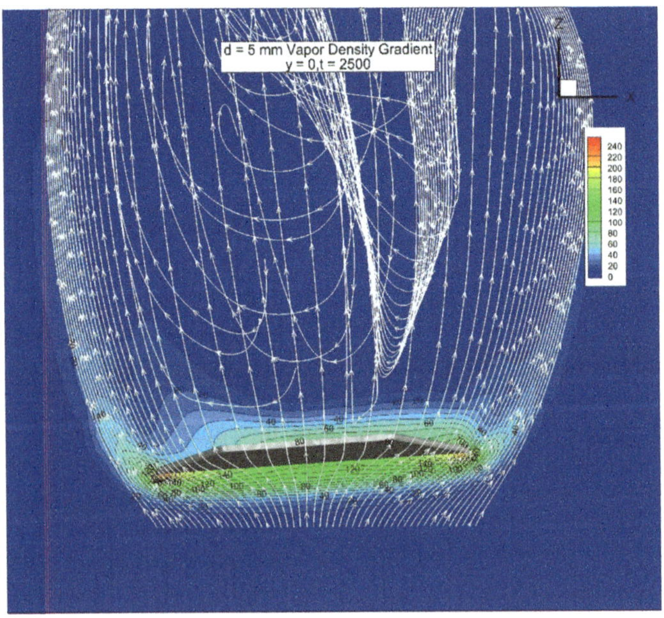

Fig. 5.5 A randomly selected snapshot of the vapor density gradient distribution around a felling hexagonal ice plate of $d = 5$ mm. The time step is the same as that in Fig. 5.4

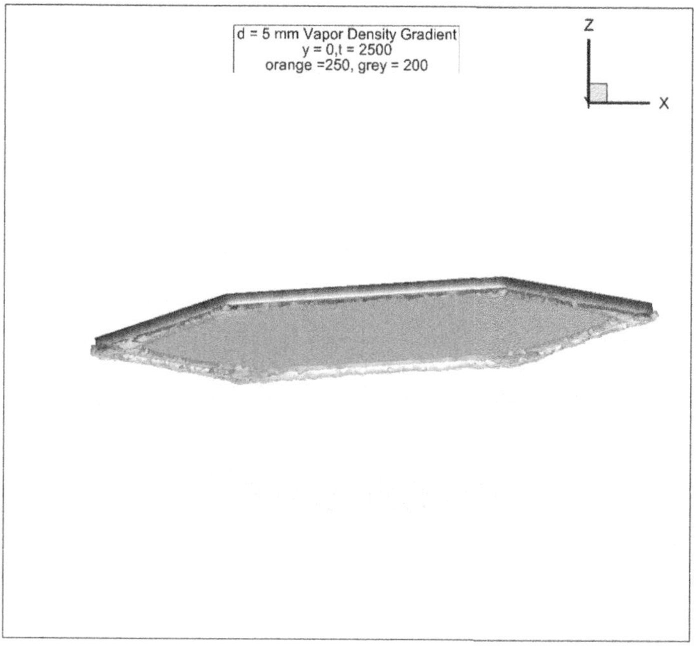

Fig. 5.6 A snapshot of the vapor density distribution around a felling hexagonal ice plate of d = 5 mm. The time step is the same as in Fig. 5.4

The above discussion on diffusion growth only considers a single factor of ice crystal growth, namely the diffusion growth rate. The habit or the shape of the crystal also depends on other factors such as temperature and environmental supersaturation as well as internal crystallographic molecular dynamics. Hence a single factor does not determine the shape of the crystal completely (see e.g., [17, Chap. 3]).

We have also computed the vapor distributions of other types of planar ice crystals as reported by Nettesheim and Wang [11]. Figures 5.7 and 5.8 show an example of the vapor distribution around a freely falling 5 mm diameter broad branch crystal (see Fig. 4.28 for the shape) in a x–z and y–z plane.

Since the shape of the crystal is more complicated this time, one may wonder where the maximum diffusion growth rate would occur. Figure 5.9 shows the y–z cross-section of the vapor density gradient revealing that the maximum gradient occurs in the underside near (but not at) the edge of each branch. Figure 5.10 shows the plot of two isosurfaces of the vapor density gradient conveying similar features of the vapor density gradient. This again implies that the edge will grow the fastest.

The results of the mean ventilation coefficient for falling planar ice crystals are summarized in Fig. 5.11, and include numerical results from Ji and Wang [7]. The results from the current study can be fit by the following empirical formulae, taking a form similar to that of Ji and Wang [7]:

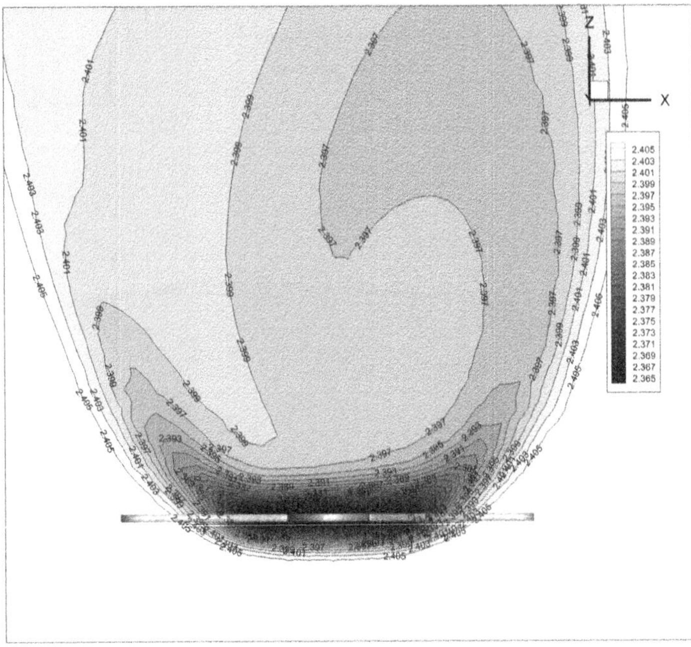

Fig. 5.7 Vapor distribution in the xz plane around a freely falling 5 mm diameter broad branch crystal

Sector plates:

$$\overline{f}_v = 1 + 0.9146Y + 3.317Y^2 - 2.127Y^3 + 0.5171Y^4 \tag{5.1}$$

Broad-branched crystals:

$$\overline{f}_v = 1 + 0.6761Y + 3.780Y^2 - 2.518Y^3 + 0.5845Y^4, \tag{5.2}$$

Stellar crystals:

$$\overline{f}_v = 1 - 0.04(X/10) + 7.92(X/10)^2 - 7.78(X/10)^3 + 7.40(X/10)^4, \tag{5.3}$$

Ordinary dendritic crystals:

$$\overline{f}_v = 1 + 0.2834Y + 6.066Y^2 - 1.612Y^3 - 0.5295Y^4, \tag{5.4}$$

where $\overline{f}_v$ is the dimensionless ventilation coefficient and $Y = X/10$ where X is a dimensionless number defined as

$$X = N_{scv}^{1/3} N_{\text{Re}}^{1/2} \tag{5.5}$$

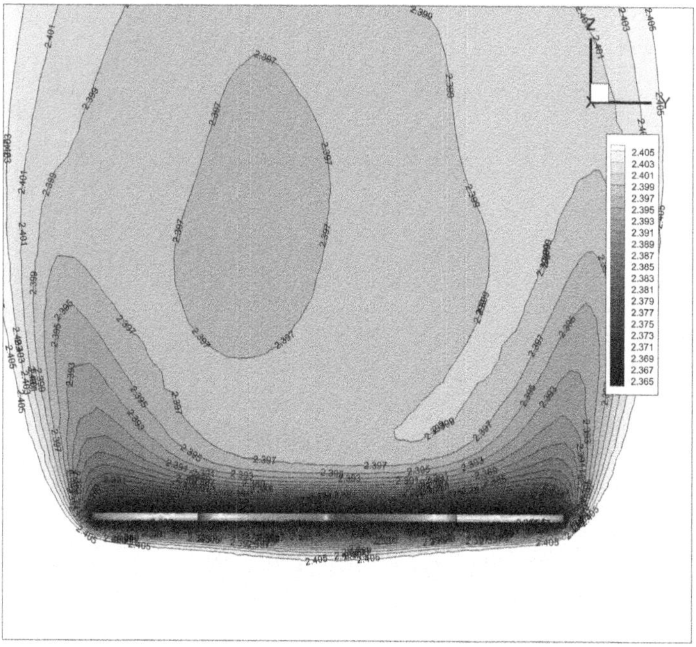

Fig. 5.8 Vapor distribution in the yz plane around a freely falling 5 mm diameter broad branch crystal

where $N_{Scv} = \nu/D_v$ is the Schmidt number of water vapor.

Pitter et al. [12] was the first one to propose a functional dependence of $\overline{f}_v$ on X. As in Ji and Wang [7], N_{Scv} is held at a constant value (= 0.63), so the relationships in Eqs. (5.1) to (5.4) are essentially between $\overline{f}_v$ and N_{Re}. The root-mean-square errors for the above relationships are 0.034, 0.018, 0.014, and 0.028, respectively, and are valid over the range of N_{Re} corresponding to each crystal type described in Table 4.4 in Chap. 4.

Figure 5.11 also shows that the ordinary dendritic crystals generally have the higher ventilation coefficient, at a given Reynolds number, compared to the other crystal habits. This becomes more pronounced with increasing N_{Re}. This can be understood by considering the dimensions and structure of the varying crystal habits. The more skeletal structure of a dendrite allows for a greater surface area that can be subjected to the ventilation effect, despite falling at a lower terminal velocity compared to the sector plate and broad-branched crystal at the same N_{Re}.

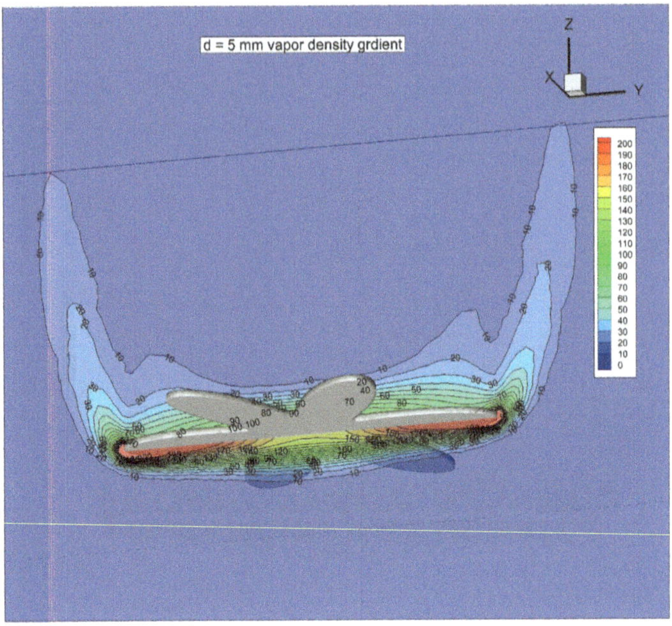

Fig. 5.9 Vapor density gradient distribution in yz plane around a freely falling 5 mm diameter broad branch crystal

5.2.2 *Vapor Density Around Falling Smooth Spherical Hailstones*

5.2.2.1 d = 1 cm case

Next we will examine the ventilation phenomenon of the hailstones. Figure 5.12 shows the computed vapor density distribution in the central ($x = 0$) cross-section of the y–z plane around a falling spherical hail of 1 cm diameter. Hailstones fall much faster than ice crystals and therefore the ventilation effect is much stronger. The total vapor flux density consists of two parts: the diffusion flux density $-\vec{u} \cdot \nabla \rho_v$ and the convection current density $\rho_v \vec{u}$. The combination of the two determines the flux pattern around the falling hydrometeor.

It is immediately clear from Fig. 5.12 that the vapor distribution is even more asymmetric than that around a falling ice crystal: the front side is surrounded by very high vapor densities whereas very lower vapor densities surround the rear. The high asymmetry is of course due to the high fall velocity of the hail so that the fore-aft asymmetry dominates the vapor distribution patter.

We see that the "driest" regions in Fig. 5.12 are on the two sides somewhat above the "equator" of the hail. In 3-dimensional view, these regions would form a continuous belt surrounding the upper part of the hail. The main reason for this dry

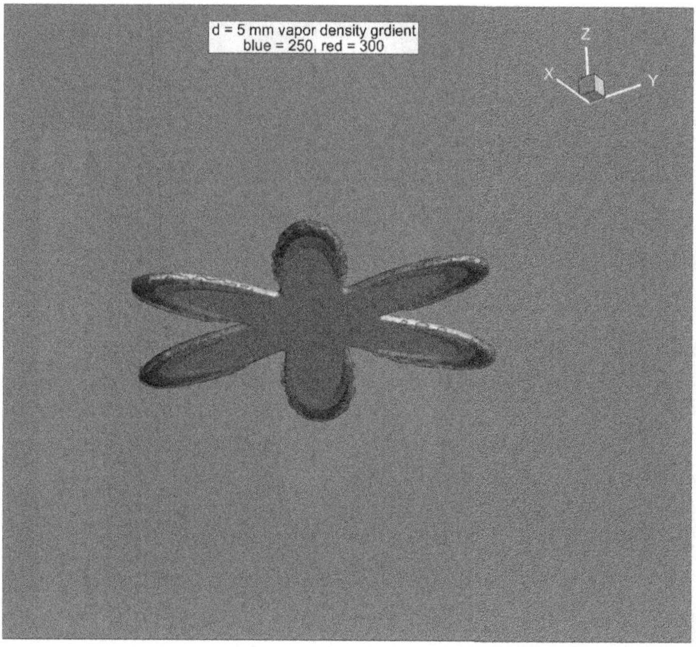

Fig. 5.10 Vapor density gradient isosurfaces around a freely falling 5 mm diameter broad branch crystal

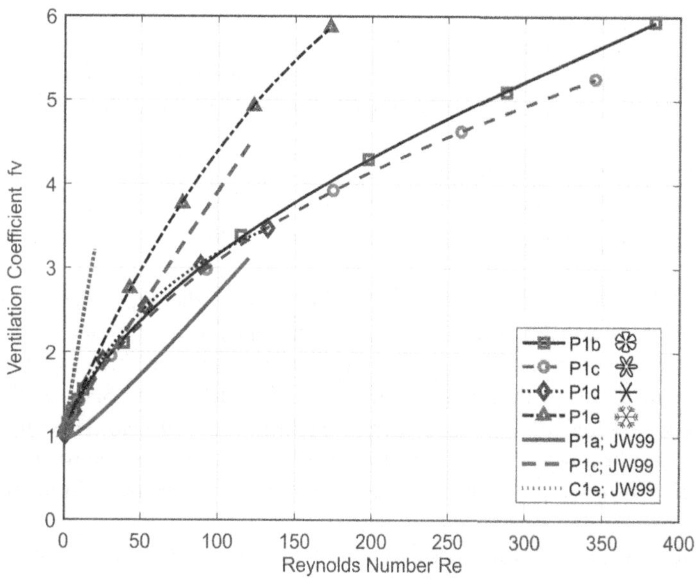

Fig. 5.11 Ventilation coefficient of falling planar ice crystals (adapted from [11] with changes)

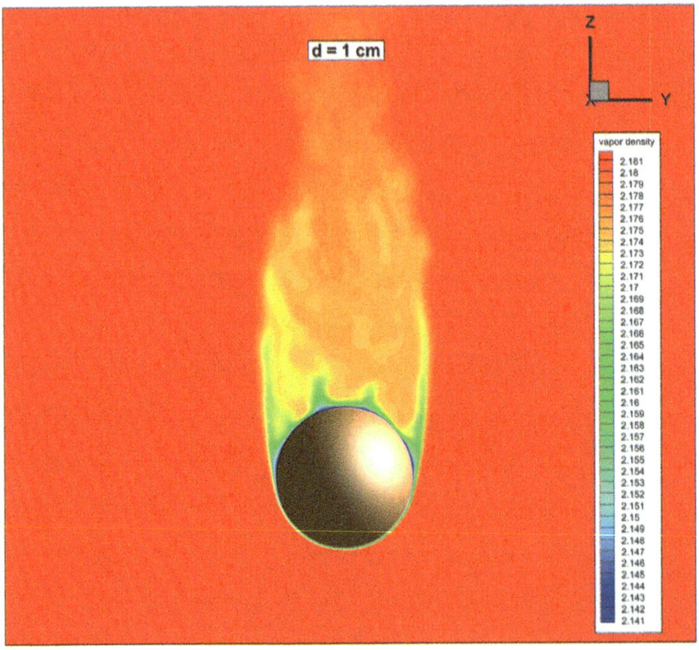

Fig. 5.12 A randomly selected snapshot of vapor density distribution around a falling spherical hailstone of $d = 1$ cm (after [5] with changes)

belt is that the air velocities outside the belt are very large which carry water vapor quickly away from the hail, leaving the region with relatively little vapor.

In the discussions of falling ice crystals, we used the magnitude of the vapor density gradient $|\nabla \rho_v|$ to illustrate the amount of vapor flux density around the hydrometeor. What really contributes to the "growth" of the hydrometeor, however, is the inward radial component of the vapor density gradient $-\nabla \rho_v \cdot \hat{e}_r$ where $\hat{e}_r$ is the unit vector in the outward radial direction, and this is what we will illustrate below.

Figure 5.13 shows the inward radial component of the vapor density gradient distribution corresponding to Fig. 5.12. Due to the high fall velocity, the enhancement of the vapor density gradient is extremely large in the upstream such that the highest gradient zone is a very thin layer immediate to the surface. We also see that there are pockets of secondary high gradient in the rear near the surface. This implies that even the rear portion of the hail can have non-negligible diffusion growth. The dry belt mentioned in the previous paragraph seems to also have the minimum inward gradient.

Note that since the flow is unsteady, the locations mentioned here change with time. But the pattern remains similar to that described here even after averaging over time.

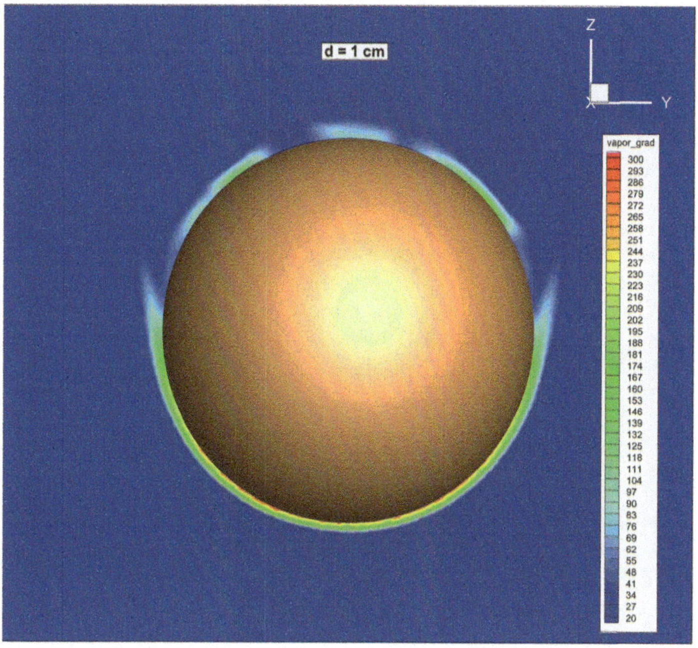

Fig. 5.13 A snapshot of vapor density gradient distribution around a falling spherical hailstone of $d = 1$ cm at the same time step as that in Fig. 5.12 (after [5] with changes)

5.2.2.2 d = 5 cm case

Next, we will use $d = 5$ cm case to illustrate the vapor distribution around larger hails. For a larger hail, the fall velocity is larger and hence the convective vapor current will dominate even more pronouncedly. Figure 5.14 shows a randomly chosen frame of the vapor density distribution in the central *yz*-plane for the case of a 5 cm diameter hail falling in air. The high vapor density contours are even closer to the front surface of the hail, implying even higher vapor density gradient, and the wake region is even drier than the case of a 1 cm hail as depicted in Fig. 5.12. This is clearly due to the much higher fall velocity of the 5 cm hail that carries the vapor around the hail by its powerful flow field. The vapor density in the wake region is strongly dispersed by the high air velocity and quickly transported downstream, resulting in the generally drier wake.

A notable feature of Fig. 5.14 is that the very dry layer near the rear surface of the hail is also clearly shallower than the corresponding layer in Fig. 5.12, despite the generally higher air velocity away from the hail. This means that the higher vapor density air can get closer to the surface of the large hail than the smaller hail case. This is because that, in a sense, eddies in the wake are counteracting the flow outside of the recirculation bubble. Because of the larger hail fall velocity, eddies developed in the bubble of the large hail possess generally stronger return flow velocities than

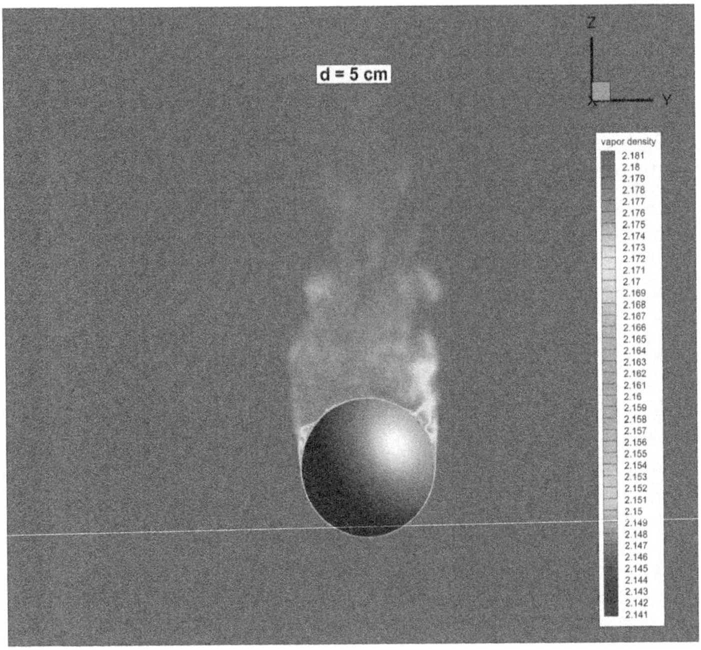

Fig. 5.14 A randomly selected snapshot of vapor density distribution around a falling spherical hailstone of $d = 5$ cm (after [5] with changes)

the case of a smaller hail. The stronger return flow can bring water vapor closer to the hail surface. An example of the return flow is shown in Fig. 5.15 to illustrate this point. Here we use the relative humidity instead of the vapor density as the humidity variable. We see that wherever the return flow velocities are larger, the high vapor density contours are brought closer to the surface. On the other hand, regions with high outward flow velocities generally have lower vapor density.

The vapor gradient distributions discussed above seem to indicate that, for pure diffusion growth, a spherical hail would grow fastest at the front end, somewhat slower at the rear, but slowest in the dry belt region. This would result in a pear-shaped hail if it continues to grow in this manner. However, to really assess the growth situation, we will need to consider the heat diffusion due to the latent heat release at the same time. Conversely, in the case of an evaporating hail, the front will evaporate the fastest whereas the dry belt mentioned above will now become a moist belt and hence evaporate the slowest. This tends to make the spherical hail to turn into a quasi-oblate spheroid. Again, this assertion is made purely on the ground of vapor diffusion. The true process needs to consider heat diffusion as well. As is well known in cloud physics community, the heat dissipation problem ultimately decides whether or not the melting of hail will commence as it falls through an atmospheric layer with temperature warmer than 0 °C.

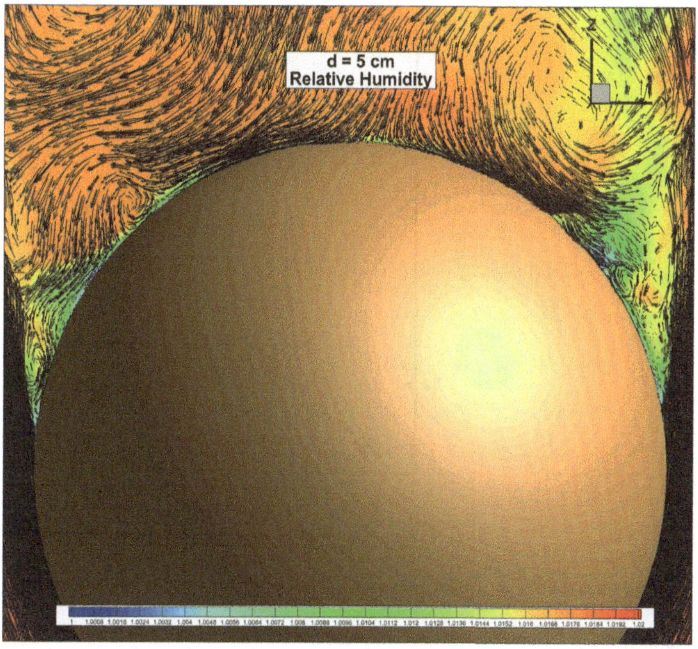

Fig. 5.15 Relative humidity with respect to ice in a central cross-section in the wake of a falling spherical hailstone of 5 cm diameter (after [5] with changes)

Based on the vapor distributions described above, we calculated the ventilation coefficients for these hailstones as described in Chap. 3 and the results are shown in Fig. 5.16.

Here the mean ventilation coefficient is plotted against the hail diameter. It is obvious that the two variables are related almost linearly and can be fitted by the following relation:

$$\overline{f}_v = 23.564d - 7.034 \tag{5.6}$$

Note that the unit of d should be in cm. A second order polynomial can fit even closer but is probably unnecessary. Sometimes it is desirable to relate the ventilation coefficient to Reynolds number instead of diameter, and this is shown in Fig. 5.17.

The curve in Fig. 5.17 can be fitted by a power relation:

$$\overline{f}_v = 0.6953 N_{\mathrm{Re}}^{-3.097} \tag{5.7}$$

A plot of the ventilation coefficients as a function of X defined in (5.5) shown in Fig. 5.18:

and the curve can be fitted by

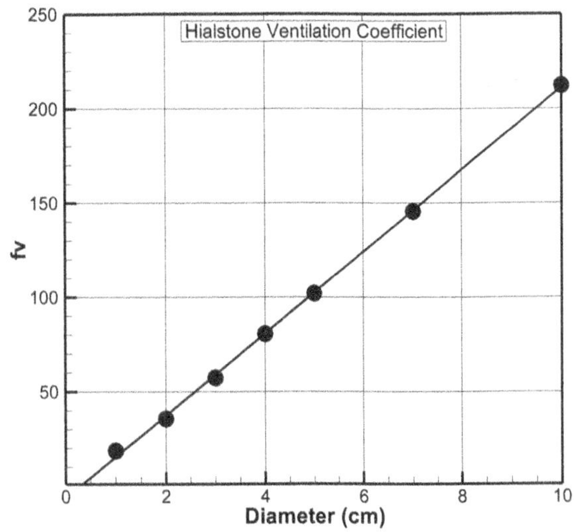

Fig. 5.16 Ventilation coefficient versus hail diameter (after [5] with changes)

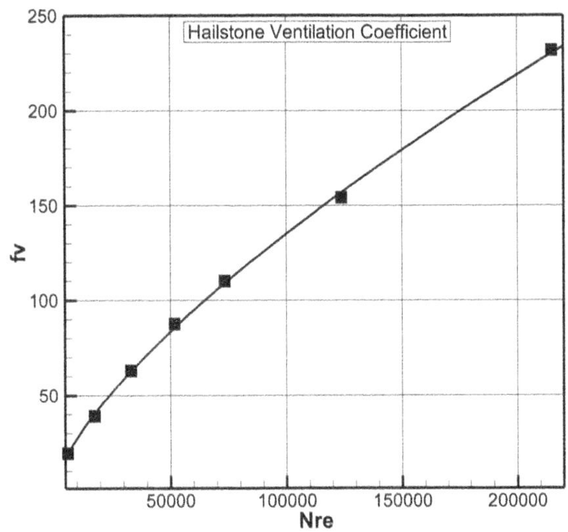

Fig. 5.17 Ventilation coefficient versus hail Reynolds number (after [5] with changes)

$$\overline{f}_v = -8.561 + 0.368X + 5.862X^2 \tag{5.8}$$

Equations (5.7) and (5.8) can be used to calculate the values of $\overline{f}_v$ for different hail sizes. Although $\overline{f}_v$ is defined for vapor diffusion, it is also often assumed that the ventilation coefficient for heat diffusion $\overline{f}_h$ is approximately the same as $\overline{f}_v$, i.e., $\overline{f}_h \simeq \overline{f}_v$ (see [13]).

Thus the ventilation effect for hails is indeed very large: the $\overline{f}_v$ for a 1 cm hail is already 20 whereas that for a 10 cm hail is over 200! One of the criteria of severe

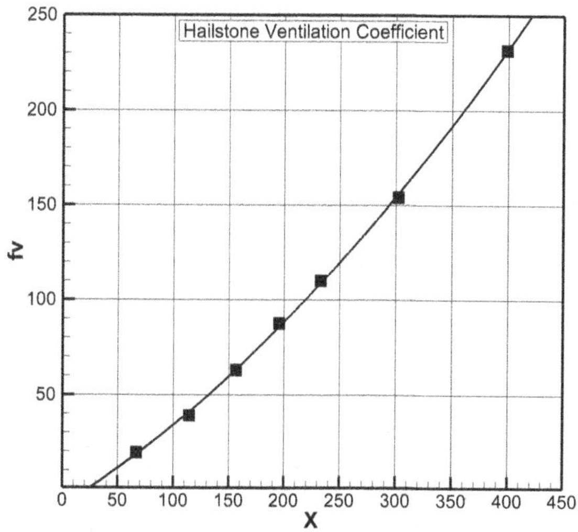

Fig. 5.18 Ventilation coefficient versus X (after [5] with changes)

thunderstorm defined by the National Weather Service is that the storm produces hails with diameter 1 inch (2.54 cm) or larger. We see that for a 2.5 cm hail the ventilation coefficient is about 50. This means that the hail when falling would grow 50 times faster than when stationary, and the same works when it is evaporating. Such rapid growth or evaporation would have significant implications to the storm within which these hails develop.

For example, one of the major concerns of a severe storm is that it may produce downburst that jeopardize aviation safety and it has been long suspected that the evaporative cooling of large hydrometeors such as large rain drops or hails is the main cause of the rapid sinking air in the downburst [6]. Considering the magnitude of the ventilation coefficients presented here, it is probably not too difficult to understand that the evaporation of large hails in a severe storm can indeed cool the air down significantly to produce high negative buoyancy. This negative buoyancy can work together with the downward drag due to the hail fall to cause severe downburst.

Similarly, in the hail growing region of the cloud, latent heat released from diffusion growth will be multiplied by such high ventilation coefficients which, when combined with the latent heat released due to riming, will produce strong updrafts and cause explosive development of that part of the cloud. It may be possible to observe such events by radar techniques. Enough riming-released latent heat may also melt the hail that renders the hail to go through wet growth regime even in a very cold environment.

5.2.3 *Vapor Distribution Around Falling Lobed Hailstones*

Next we present the results for lobed hailstones which were reported in detail by Wang and Chueh [18]. In the following, we will show the calculated vapor density distributions and ventilation coefficients for the 6 cases of lobed hailstones—3 long lobed and 3 short lobed. As we shall see, the irregularities on the surface complicate the ventilation behavior greatly so that the ventilation effect is not simply a matter of size or velocity or the number of lobes.

5.2.3.1 Short Lobed Cases

Figure 5.19 shows the computed vapor density distribution in the central cross-section of the yz-plane around the three cases of short lobed hailstones of 1 cm diameter. All the front (upstream) sides of the three lobed hailstones are surrounded by high vapor densities, while low vapor densities surround their rear sides, implying that the convective flux $\rho_v\vec{u}$ is generally dominant over the diffusion flux $-D_v\nabla\rho_v$ due to the motion of the hailstones (i.e. the fluid inertial effect on the vapor transport is much greater than the diffusion of water vapor effect.) such that the vapor convective flux $\rho_v\vec{u}$ in the front points toward the hailstone surface, whereas the flux in the rear is directed away from the surface, which is consistent with the results of spherical hailstones described in the last section.

In general, there is no much difference among these three short lobed hailstones in terms of water vapor distribution (see Fig. 5.19). The distribution in the front side of the 3 cases are similar, whereas that in the rear show more differences. The drier

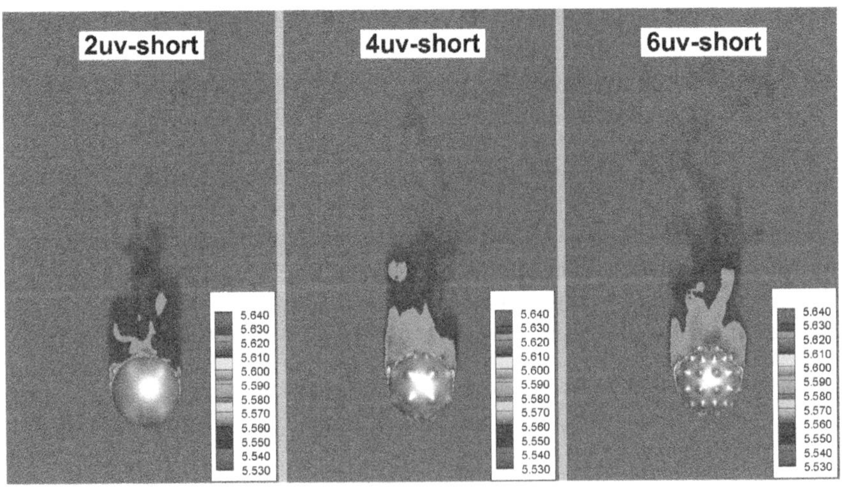

Fig. 5.19 Vapor density distribution in the central vertical cross-section of short-lobed hailstones of $d = 1$ cm at randomly chosen time frames (after [18] with changes)

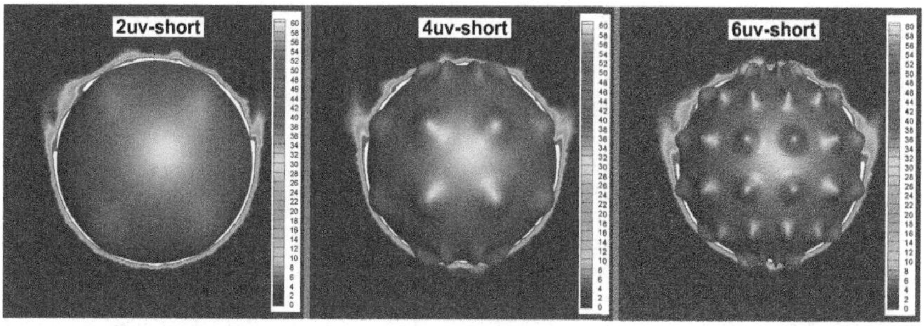

Fig. 5.20 Vapor density gradient distribution in the central vertical cross-section of short-lobed hailstones of $d = 1$ cm at time frames corresponding to that in Fig. 5.19 (after [18] with changes)

regions in the rear near the hailstone surface are somewhat larger and more irregular with increasing number of lobes. This should indicate the formation of more turbulent wakes with increasing number of lobes which prevents higher vapor content air to get close to the hail surface easily.

To further understand the diffusion growth of the lobbed hailstone, we plot the vapor gradient magnitude $|\nabla \rho_v|$ as shown by Fig. 5.20.

It is seen here that in all three cases the front side is characterized by high vapor gradient near the surface while the rear side is low vapor gradient, as they should be. Among the three, however, the 2uv-short case has a nearly hemispherical coverage of high gradient in the front side whereas the coverage becomes smaller and smaller as the number of lobes increase. In the rear side, the 2uv-short case shows three isolated pockets of high gradient areas whereas the number of such pockets also becomes fewer as the number of lobes increases. A look into the microstructure of the flow field near the rear surface of the hailstone (not shown) reveals that the 2uv-short case has the largest and most organized inflow regions that bring in water vapor towards the surface whereas these inflow regions are smaller and most disorganized for the 6uv-short case. Thus it appears that the irregularities or roughness on the surface hinders the flux of water vapor towards the hailstone. Similar conclusion has been given by Aguirre Varela et al. [1] who studies similar process for graupel and found that the rear area with turbulent wakes has less heat/mass extraction than the front area. But, at the same time, we note that the surface areas of the 4uv- and 6uv-cases are greater that the 2uv case, so that the increase in area compensates the smaller gradient that results in the very similar ventilation coefficients (under the assumption that all three have the same Reynolds number) as we shall see later.

Figure 5.21 shows a randomly chosen frame of the vapor density distribution in the central yz-plane for the case of 5 cm diameter hailstone falling in air. Here we again see that the dry area (the yellow region) in the rear of the hailstone is most extensive in the case of 6uv-short and the thinnest in the 2uv-short case. Thus, one would expect that the vapor flux density towards the surface is the weakest for the 6uv-short case. This is confirmed by Fig. 5.22, which shows the vapor density gradient around the

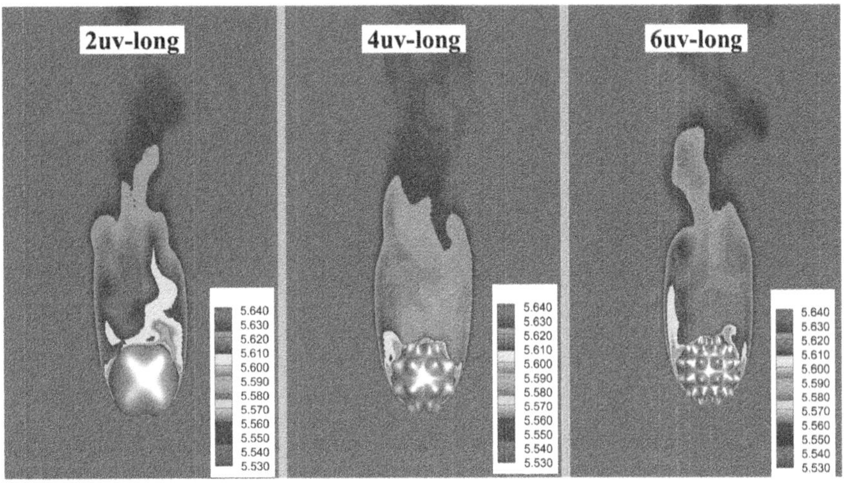

Fig. 5.21 Vapor density distribution in the central vertical cross-section of short-lobed hailstones of $d = 5$ cm at randomly chosen time frames (after [18] with changes)

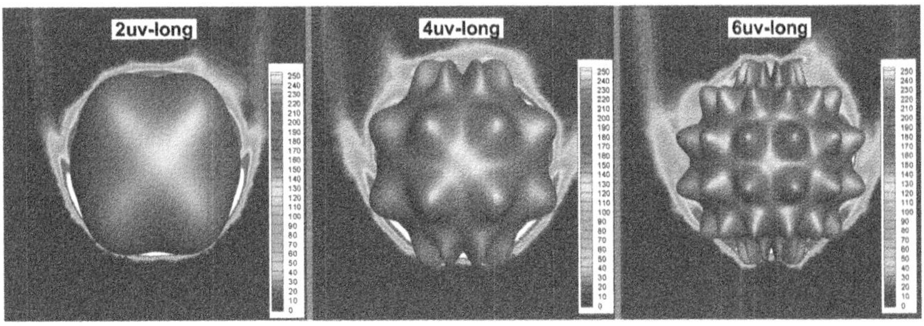

Fig. 5.22 Vapor density gradient distribution in the central vertical cross-section of short-lobed hailstones of $d = 5$ cm at time frames corresponding to that in Fig. 5.21 (after [18] with changes)

hailstones where the high gradient areas appear to be the smallest in the 6uv-short case.

Figure 5.23 shows the calculated ventilation coefficients $\overline{f}_v$ of the short lobed hailstones of diameters 1, 3, 5, 7 and 10 cm. It is seen that the relation of $\overline{f}_v$ versus d is not very simple. For d smaller than ~7 cm, $\overline{f}_v$ of 4uv- and 6uv-short cases are similar and smaller than that of the 2uv-short cases. For $d > 7$ cm, $\overline{f}_v$ of 4uv- and 6uv-short cases become slightly greater than that of the 2uv-short cases. Apparently, for smaller short-lobed hailstones, the number of lobes seems to hinder the vapor transport towards the surface despite that the surface areas of the 4uv and 6uv cases are greater. On the other hand, for large stones the opposite is true. The exact reason requires more study to clarify. It is possible that the larger size causes the convective

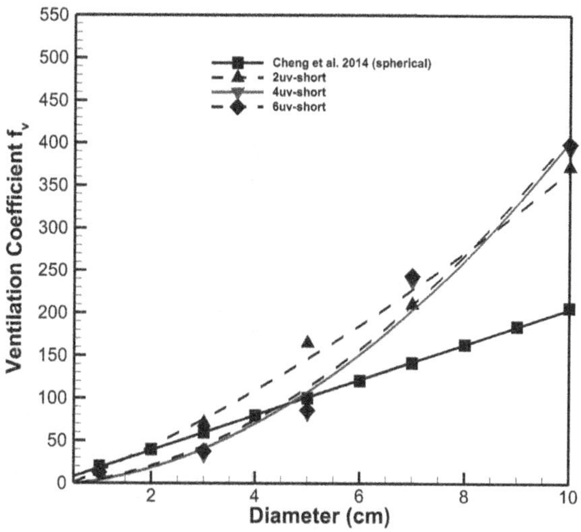

Fig. 5.23 Ventilation coefficient versus diameter for short lobed and spherical hailstones (after [18] with changes)

transport term to increase, and the integration of increased $\nabla \rho_v$ over the slightly larger surface areas of the 4uv and 6uv stones results in larger $\overline{f}_v$.

Figure 5.23 also shows the $\overline{f}_v$ of spherical hailstones calculated by Cheng et al. [5]. First of all, we note that the $\overline{f}_v$ of the 2uv case is, except for $d = 1$ cm, greater than that of a sphere of the same d. For 4uv and 6uv cases, the $\overline{f}_v$ is smaller than that of a sphere of the same diameter if $d < 5$ cm and becomes larger if $d > 5$ cm. It is seen that for larger hailstones, the lobes help to significantly enhance the ventilation effect.

The curves in Fig. 5.23 can be fitted by the following empirical expressions:

$$\overline{f}_v = 20.66d - 2.259 \quad \text{for sphere} \tag{5.9}$$

$$\overline{f}_v = 0.3086d^3 - 3.827d^2 + 47.86d - 34.29 \quad \text{2uv-short} \tag{5.10}$$

$$\overline{f}_v = 3.687d^2 + 3.713d - 3.569 \quad \text{4uv-short} \tag{5.11}$$

$$\overline{f}_v = 3.589d^2 + 5.42d - 4.416 \quad \text{6uv-short} \tag{5.12}$$

Sometimes it is desirable to see the relation between the ventilation coefficient and the Reynolds number instead of the diameter d, and this is shown in Fig. 5.24.

The fitting equations for curves using $Y = \log_{10} N_{\mathrm{Re}}$ in Fig. 5.24 are as follows:

$$\overline{f}_v = \exp(6.867Y - 6.545) \quad \text{for sphere} \tag{5.13}$$

$$\overline{f}_v = \exp(9.783Y - 10.92) \quad \text{2uv-short} \tag{5.14}$$

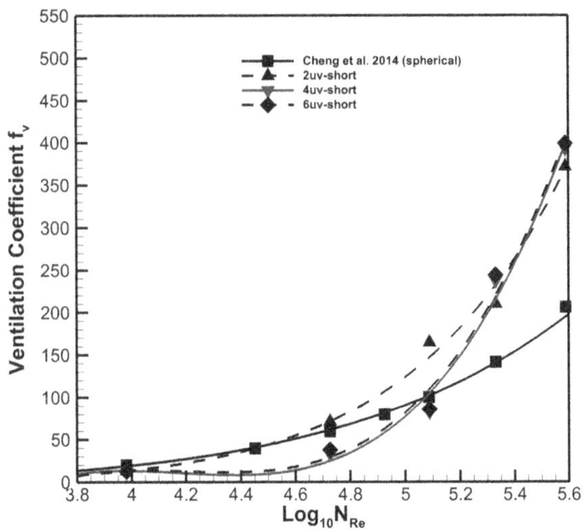

Fig. 5.24 Ventilation coefficient versus $\log_{10} N_{Re}$ for short lobed and spherical hailstones (after [18] with changes)

$$\overline{f}_v = 151.9Y^3 - 1914Y^2 + 8025Y - 11190 \quad \text{4uv-short} \tag{5.15}$$

$$\overline{f}_v = 151.9Y^3 - 1914Y^2 + 8025Y - 11190 \quad \text{6uv-short} \tag{5.16}$$

Equations (5.15) and (5.16) are identical because the two curves are nearly on top of each other and the polynomial fits of $\overline{f}_v(Y)$ for the two cases yield the same coefficients.

Figure 5.25 shows the plot of $\overline{f}_v$ versus X and the empirical fit relations including an empirical relation suggested by Rasmussen and Heymsfield [15] are:

$$\overline{f}_v = 0.0003137X^2 + 0.386X - 6.483 \quad \text{sphere} \tag{5.17}$$

$$\overline{f}_v = 0.00124X^2 + 0.4229X - 18.38 \quad \text{2uv-short} \tag{5.18}$$

$$\overline{f}_v = 0.003014X^2 - 0.2994X + 13.19 \quad \text{4uv-short} \tag{5.19}$$

$$\overline{f}_v = 0.002953X^2 - 0.2538X + 11.13 \quad \text{6uv-short} \tag{5.20}$$

$$\overline{f}_v = 0.78 + 0.308X \quad \text{(Rasmussen and Heymsfield 1987)} \tag{5.21}$$

Figure 5.25 shows that the estimate of ventilation effect may differ greatly if different fit equations are adopted in cloud models, especially for severe storms when very large hailstones are produced. For storm cases involving only small hailstones (d < 3 cm) the differences would be small.

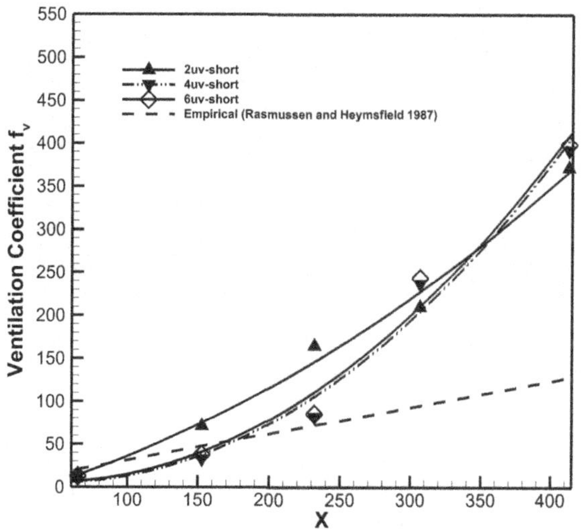

Fig. 5.25 Ventilation coefficient versus X for short lobed and spherical hailstones (after [18] with changes)

5.2.3.2 Long Lobed Cases

The results of the long lobed hailstones are shown in this section and we can make a comparison between these hailstones and the short ones present in the previous section. Figure 5.26 shows the randomly selected vapor distribution of 1 cm three long lobed cases.

It is seen in this figure that the drier region is generally larger than those of the short lobed cases shown in Fig. 5.19. Hence it appears that the downstream dry

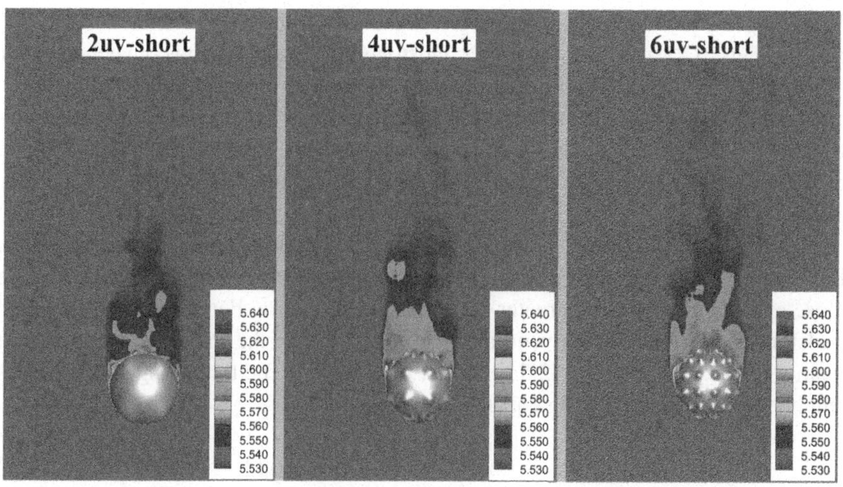

Fig. 5.26 Vapor density distribution in the central vertical cross-section of long-lobed hailstones of $d = 1$ cm at randomly chosen time frames (after [18] with changes)

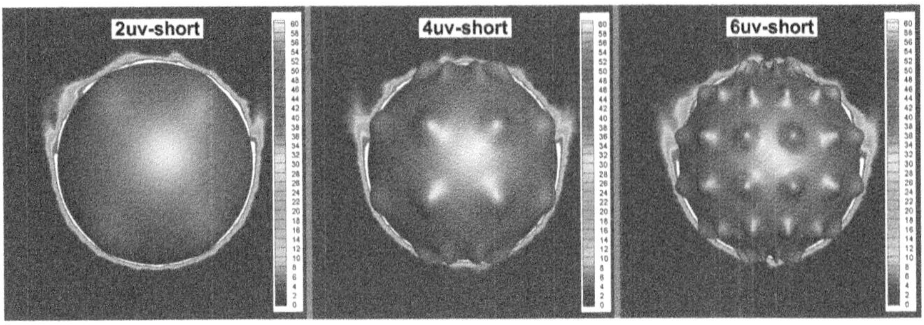

Fig. 5.27 Vapor density gradient distribution in the central vertical cross-section of short-lobed hailstones of $d = 1$ cm corresponding to the time frames in Fig. 5.26 (after [18] with changes)

region tends to become greater as the lobe becomes longer. On the other hand, the dry region doesn't always become greater as the number of protrusions rises, since the size of the 4uv green region is smaller than those of the other cases and the 6uv green region appears to be smaller than that of the 2uv case. The situation is complicated undoubtedly due to the complex shape and the exact reason requires further study.

Figure 5.27 shows the vapor gradient magnitude distribution of 1 cm long lobed hailstones. Compared with those of the corresponding short lobed cases shown in the previous section, the high gradient areas become smaller due to the presence of more turbulent regions between the long lobes that hinder the fluid flow onto the surface. In addition, as the number of protrusions increases, the high gradient region becomes smaller as it becomes harder for water vapor to enter the spherical part of the hailstone.

Figures 5.28 and 5.29 show the water vapor and vapor gradient distributions of 5 cm long lobed hailstones respectively. Again, as in the case of short lobed 5 cm cases in Figs. 5.21 and 5.22, the larger hailstone falls faster and the convective vapor flux is strong enough to dominate over the diffusion vapor flux. The high vapor density region now can get closer to spherical surface than does the corresponding short-lobed one. It is also clear that as the hailstone becomes larger and falls faster, the boundary layer becomes thinner permitting the higher density vapor to get closer to the surface.

Figure 5.30 shows the calculated ventilation coefficients for long lobed hailstones. For smaller hailstones with $d < 3$ cm the ventilation coefficients of three cases are close and are only slightly greater than the spherical case. For $d > 3$ cm, all lobed hailstones have much higher ventilation coefficients than the spherical ones, ranging from ~1.5 for $d = 4$ cm to ~2.5 times for $d = 10$ cm. The three long-lobed curves can be fitted by the following equations:

$$\overline{f}_v = 0.2819d^3 - 3.543d^2 + 49.63d - 37.33 \quad \text{2uv-long} \tag{5.22}$$

$$\overline{f}_v = -0.5187d^3 + 10.3d^2 - 7.439d + 10.94 \quad \text{4uv-long} \tag{5.23}$$

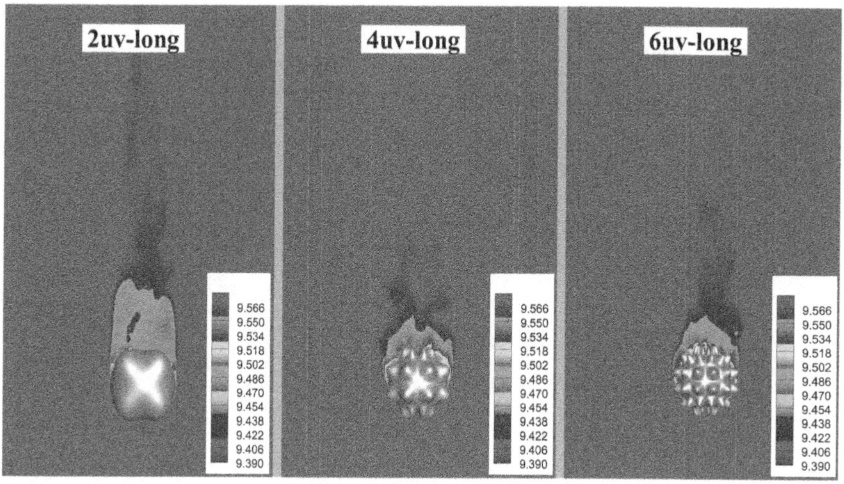

Fig. 5.28 Vapor density distribution in the central vertical cross-section of long-lobed hailstones of $d = 5$ cm at randomly chosen time frames (after [18] with changes)

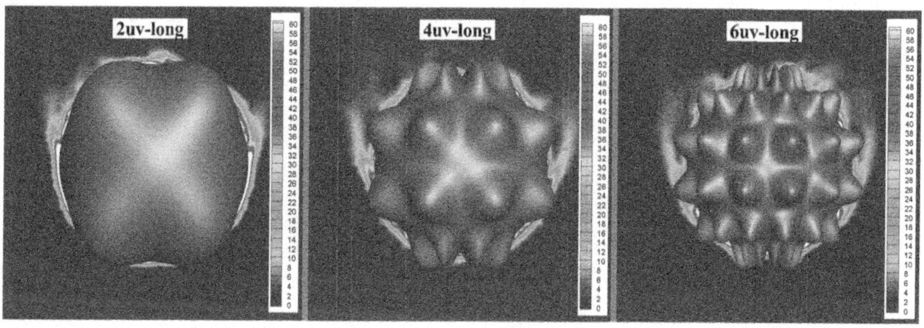

Fig. 5.29 Vapor density gradient distribution in the central vertical cross-section of long-lobed hailstones of $d = 5$ cm at the time frames corresponding to that in Fig. 5.28 (after [18] with changes)

$$\overline{f}_v = -0.3016d^3 + 7.277d^2 + 9.418d - 1.024 \quad \text{6uv-long} \tag{5.24}$$

Figure 5.31 shows the ventilation coefficients versus Reynolds number for long-lobed hailstones, and (5.25)–(5.27) show the corresponding curve fittings:

$$\overline{f}_v = \exp(10.08Y - 11.36) \quad \text{2uv-long} \tag{5.25}$$

$$\overline{f}_v = \exp(10.61Y - 12.15) \quad \text{4uv-long} \tag{5.26}$$

$$\overline{f}_v = \exp(10.34Y - 11.56) \quad \text{6uv-long} \tag{5.27}$$

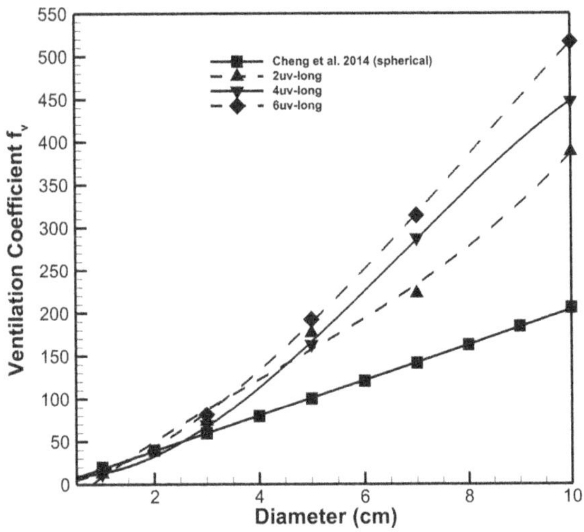

Fig. 5.30 Ventilation coefficient versus diameter for long lobed and spherical hailstones (after [18] with changes)

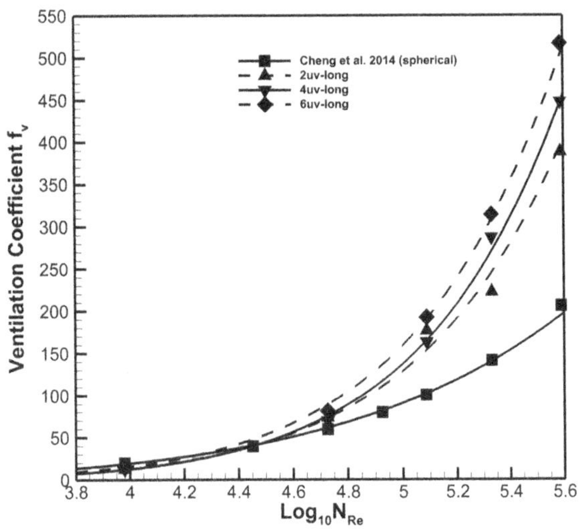

Fig. 5.31 Ventilation coefficient versus $\log_{10} N_{\mathrm{Re}}$ for long lobed and spherical hailstones (after [18] with changes)

Again $Y = \log_{10} N_{\mathrm{Re}}$.

Figure 5.32 shows the ventilation coefficients versus the non-dimensional X for long-lobed hailstones, and (5.28)–(5.30) show the corresponding curve fittings:

$$\overline{f}_v = 4.103 \times 10^{-6} X^3 - 1.778 \times 10^{-3} X^2 + 1.106X - 55.64 \quad \text{2uv-long} \tag{5.28}$$

$$\overline{f}_v = -8.221 \times 10^{-6} X^3 + 7.648 \times 10^{-3} X^2 - 0.7469X + 31.79 \quad \text{4uv-long} \tag{5.29}$$

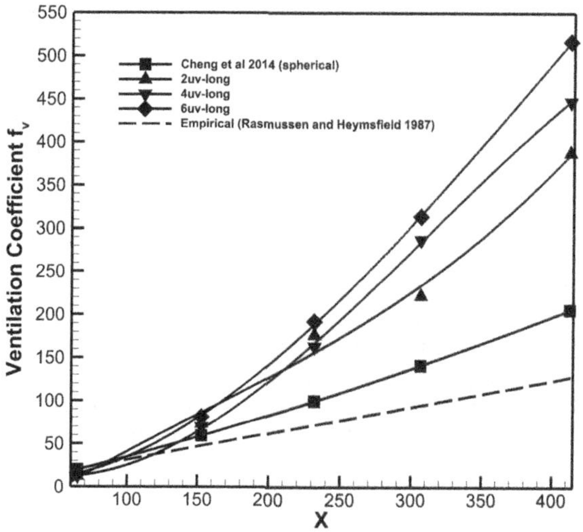

Fig. 5.32 Ventilation coefficient versus X for long lobed and spherical hailstones (after [18] with changes)

$$\overline{f}_v = -4.589 \times 10^{-6}X^3 + 5.47 \times 10^{-3}X^2 - 0.2463X + 9.533 \quad \text{6uv-long} \tag{5.30}$$

Although the $\overline{f}_v$ reported above is evaluated in terms of water vapor diffusion, it is usually assumed that the ventilation coefficient for heat diffusion $\overline{f}_h$ is the same as $\overline{f}_v$ [13].

References

1. Aguirre Varela GG, Castellano NE, Pereyra RG, Avila EE (2003) The effect of surface lobes on heat transfer from an ice particle. Q J R Meteor Soc 129:3425–3438
2. Bailey IH, Macklin WC (1968) The surface configuration and internal structure of artificial hailstones. Q J R Meteor Soc 94:1–11
3. Beard KV, Pruppacher HR (1971) A wind tunnel investigation of the rate of evaporation of small water drops falling at terminal velocity in air. J Atmos Sci 28:1455–1464
4. Bentley WA, Humphreys WJ (1962) Snow crystals. Dover, 226 pp
5. Cheng KY, Wang PK, Wang CK (2014) A Numerical study on the ventilation coefficients of falling hailstones. J Atmos Sci 71:2625–2634
6. Fujita TT (1978) Manual of downburst identification for Project Nimrod. SMRP research paper. Satellite and mesometeorology research project. Dept. of the Geophysical Sciences, University of Chicago, Chicago, 104 pp
7. Ji W, Wang PK (1999) Ventilation coefficients for falling ice crystals in the atmosphere at low-intermediate Reynolds numbers. J Atmos Sci 56:829–836
8. List R (1963) General heat and mass exchanges of spherical hailstones. J Atmos Sci 20:189–197
9. Macklin WC (1963) Heat transfer from hailstones. Q J R Meteor Soc 89:360–369
10. Masliyah JH, Epstein N (1970) Numerical study of steady flow past spheroids. J Fluid Mech 44:493–512
11. Nettesheim J, Wang PK (2018) A numerical study on the aerodynamics of freely falling planar ice crystals. J Atmos Sci 75:2849–2865

12. Pitter RL, Pruppacher HR, Hamielec AE (1974) A numerical study of the effect of forced convection on mass transport from a thin oblate spheroid of ice in air. J Atmos Sci 31:1058–1066
13. Pruppacher HR, Klett JD (1997) Microphysics of clouds and precipitation, 2nd edn. D. Reidel, New York, 954 pp
14. Pruppacher HR, Rasmussen RM (1979) A wind tunnel investigation of the rate of evaporation of large water drops falling at terminal velocity in air. J Atmos Sci 36:1255–1260
15. Rasmussen RM, Heymsfield AJ (1987) Melting and shedding of graupel and hail. Part I: model physics. J Atmos Sci 44:2754–2763
16. Thorpe AD, Mason BJ (1966) The evaporation of ice spheres and ice crystals. Br J Appl Phys 17:541
17. Wang PK (2013) Physics and dynamics of clouds and precipitation. Cambridge University Press, 467 pp
18. Wang PK, Chueh CC (2020) A numerical study on the ventilation coefficients of falling lobed hailstones. Atmos Res 234:104737
19. Woo SE, Hamielec AE (1971) A numerical method of determining the rate of evaporation of small water drops falling at terminal velocity in air. J Atmos Sci 28:1448–1454